D0426804

AIR

Our Planet's Ailing Atmosphere

Hans Tammemagi

OXFORD
UNIVERSITY PRESS

OXFORD
UNIVERSITY PRESS

70 Wynford Drive, Don Mills, Ontario M3C 1J9
www.oupcanada.com

Oxford University Press is a department of the University of Oxford.
It furthers the University's objective of excellence in research, scholarship,
and education by publishing worldwide in

Oxford   New York
Auckland   Cape Town   Dar es Salaam   Hong Kong   Karachi
Kuala Lumpur   Madrid   Melbourne   Mexico City   Nairobi
New Delhi   Shanghai   Taipei   Toronto

With offices in
Argentina   Austria   Brazil   Chile   Czech Republic   France   Greece
Guatemala   Hungary   Italy   Japan   Poland   Portugal   Singapore
South Korea   Switzerland   Thailand   Turkey   Ukraine   Vietnam

Oxford is a trade mark of Oxford University Press
in the UK and in certain other countries

Published in Canada
by Oxford University Press

Library and Archives Canada Cataloguing in Publication

Tammemagi, H. Y.
Air : our planet's ailing atmosphere / Hans Tammemagi.

Includes bibliographical references and index.
ISBN 978-0-19-543007-3

1. Air. 2. Atmosphere. 3. Air—Pollution. I. Title.
QC861.3.T34 2009   551.51   C2008-907065-8

Cover image: Prill Mendien Design & Fotografie/iStockphoto

1 2 3 4 - 12 11 10 09

The production of the title **Our Endangered Air** on Rolland Enviro 100 Print paper
instead of virgin fibres paper reduces your ecological footprint by :

Tree(s) : 12   Suspended particles in the water : 2.3 kg
Solid waste : 357 kg   Air emissions : 784 kg
Water : 33,791 L   Natural gas : 51 m³

FSC
Mixed Sources
Product group from well-managed
forests, controlled sources and
recycled wood or fiber
Cert no. SW-COC-000952
www.fsc.org
© 1996 Forest Stewardship Council

Printed by Transcontinental. Text pages on Rolland Enviro 100, containing 100% post-consumer recycled fibers,
Eco-Logo certified, processed without chlorinate, FSC Recycled and manufactured using biogaz energy.

Contents

Acknowledgements

Many people have contributed and offered encouragement during the long journey (over six years) in making this book become reality. I am grateful to the following: early drafts were improved by the reviews of Dr. Michael Waddington, McMaster University, and Dr. Duane Pendergast, Computare Ltd. Dr. Jennie Rubio at Oxford University Press patiently, enthusiastically, and expertly edited the rough edges from the text with the assistance of Robert Gardner and Leslie Saffrey. Dr. David Rodenhuis, University of Victoria, and Dr. Tom McElroy at the Meteorological Service of Canada, Environment Canada, Downsview, provided detailed review and stimulating discussion. Paul Zizek provided research assistance. Once again I am indebted to my wife, Allyson, for her loving support throughout this lengthy undertaking.

The following individuals and organizations are thanked for providing helpful discussion, review, information about their facilities or equipment, and/or giving permission to use their photographs in this book: R. Turle, Environmental Technology Centre, Environment Canada, Ottawa; D. Yap, Ontario Ministry of Environment, Toronto; Dr. H.-J. Ziock of Los Alamos National Laboratory, New Mexico; Heather Mackey, Dr. Tom McElroy, Dr. Pierrette Blanchard, Henry Hengeveld, David Phillips at the Meteorological Service of Canada, Environment Canada, Downsview; Matthew Heverly of Met One Instruments, Grants Pass, Oregon; Doug Pollard of the Federation of Canadian Municipalities, Ottawa.

Credit for photographs is given to the appropriate organization in the figure captions. Copyright remains with the person or organization credited for the photograph.

Glossary

Biosphere: the land, air, and water where living beings reside.

Buffering capacity: the ability to neutralize an acid.

Catabatic wind: warm winds. The air becomes warmer and drier as it descends a mountainside and the atmospheric pressure increases; the increasing temperature lowers relative humidity.

CFCs: chlorofluorocarbons; synthetic chemicals used in refrigeration.

Chinook wind: a catabatic wind that blows down the eastern side of the Rocky Mountains.

Cogeneration: a method of producing electricity in which "waste" heat is put to a useful purpose (rather than dissipated into the environment) such as heating homes.

Condensation: the process in which a gas changes to a liquid.

Convection: the transfer of heat by a circulatory motion in a gas or liquid caused by density differences that are, in turn, caused by temperature differences.

Coriolis force: caused by the Earth's rotation, it causes freely moving bodies such as wind and ocean currents to veer to the right (left) in the northern (southern) hemisphere.

Cosmic rays: energetic particles originating from space. Almost 90 percent are protons, about 9 percent are helium nuclei (alpha particles), and about 1 percent are electrons. The term "ray" is a misnomer, as cosmic rays are actually particles.

Electrolysis: the decomposition of water into hydrogen and oxygen by passing an electric current through it.

Foehn wind: a catabatic wind that blows down the southern side of the Alps in Europe.

Half-life: the time required for a quantity of a radioactive element to decay to half the quantity.

Halocarbons: chemical compounds in which one or more carbon atoms are linked by sharing electrons with one or more of the halogen atoms: fluorine, chlorine, bromine, or iodine.

Heat capacity: the quantity of heat required to raise the temperature of a body by one degree.

Ion: an atom, molecule, or molecular fragment carrying a positive or negative electrical charge.

Isotopes: atoms of the same element with the same number of protons but with different numbers of neutrons in the nucleus.

Methane: a colorless, odorless, flammable gas with the formula CH_4. It is the main constituent of natural gas.

Milankovitch cycles: the variations in the amount of solar radiation reaching Earth because of changes in: (a) the elliptical shape of the Earth's orbit around the sun—a 90,000- to 100,000-year cycle; (b) the tilt of the Earth's axis—a 40,000-year cycle; and (c) the precession (a slow movement of the axis of rotation of a spinning body about another line intersecting it) of the Earth's axis—a 21,000-year cycle.

Particulate matter (PM), or particulates: a complex mixture of extremely small particles and liquid droplets suspended in air.

Planetisimals: kilometer-sized objects that formed from dust during the early evolution of the solar system.

Precession: a slow movement of the axis of rotation of a spinning body about another line intersecting it.

Santa Ana wind: a catabatic wind that blows down the southwest slope of the coastal ranges of southern California.

Sequestration: the permanent disposal of carbon dioxide deep underground.

Squall line: a band of stormy weather with squally winds accompanying a cold front.

Supercooled water: this occurs when water is chilled below its freezing point without it becoming solid. A liquid below its freezing point will crystallize in the presence of a seed crystal or nucleus. Lacking any such nucleus, the liquid can be maintained below the freezing temperature.

Thermohaline conveyer belt: a worldwide system of ocean currents driven by differences in temperature and salinity (that is, density).

Trade winds: westward blowing winds between the equator and latitudes 25 to 35 degrees.

Units

Nanometer: one thousand millionth (10^{-9}) of a meter.

Microgram: one millionth (10^{-6}) of a gram.

Micrometer: one millionth (10^{-6}) of a meter. Also known as a micron.

Micron: see Micrometer.

MicroSievert: one millionth of a Sievert, where a Sievert is a unit that measures the health risk of exposure of a living organism to ionizing radiation.

Parts per million (ppm): one part in a million.

Millibar (mb): one thousandth of a bar, where a bar is a unit of pressure equal to 100,000 pascals.

Pascal: a unit of pressure that equals a force of one newton per square meter. A newton is the force that would impart an acceleration of one meter per second every second to a mass of one kilogram.

Scientific notation: also known as exponential notation, is a way of writing numbers that are too large or small to be conveniently written in standard decimal notation. In the number 10^5, the superscript 5 is the exponent and means that 10 will be multiplied by itself five times, that is, $10^5 = 10 \times 10 \times 10 \times 10 \times 10 = 100,000$. An example: $3.7 \times 10^3 = 3,700$. If the exponent is negative, say, 4^{-3}, then the number 4 is divided into 1 three times, that is, $4^{-3} = ¼ \times ¼ \times ¼$. Another example, $3.7 \times 10^{-2} = 0.037$.

Ton: 2,000 pounds.

Tonne: 1,000 kilograms (2,200 pounds), also known as the metric ton.

Figures

Preface

The thin, wispy collection of gases that we call the atmosphere makes our blue planet habitable and comfortable, and is essential to every living thing in the world. Yet it is coming under attack.

The exponential growth in human population—there are almost seven billion of us—combined with an increasing appetite for material wealth, has led to an explosion in consumer goods including four-wheel drive cars, personal computers, cell phones, brand-name clothes, motor-driven corkscrews, giant flat-screen televisions, and the list goes on and on. In North America, in particular, there are more and more of us, all clamoring for material goods.

But what is the price? After all, we all know there is no "free lunch". To manufacture all these consumer goods, not to mention essentials like homes, highways, hospitals, subways, and schools, requires energy and mountains of raw resources—some of which, like oil and natural gas, are now dwindling. And it also requires the increasing use of synthetic chemicals—many of which we discover only much later are harmful to the biosphere. Unfortunately, in the last few decades many symptoms have appeared that indicate the environment is wilting *on a global scale* under the twin onslaughts of population growth and rampant consumerism.

This book focuses on one specific environmental aspect: air. Not that water pollution, loss of species, municipal wastes, or nuclear wastes are unimportant. Because the atmosphere is fragile and sensitive to the attack of pollutants, it is an excellent yardstick for measuring the health of the overall globe. In many ways the atmosphere, the air we breathe, is the canary in the coal miner's cage.

And the canary is choking. We are facing one major problem after another, and each new one is more complex, more threatening than the previous. Air pollution has spread from local in scale to regional; and in recent decades, it has become a global problem. Contaminants can drift with the prevailing air currents for hundreds and even thousands of kilometers, despoiling pristine areas far away.

Alarms have been raised, but the media are focused almost exclusively on global warming and climate change. Headlines constantly blare that glaciers are shrinking, polar bears are in peril, sea levels are rising, and hurricanes are increasing in frequency and intensity. Lost in these info-bites is the fact that a changing climate is only the tip of the iceberg toward which our global *Titanic* is steaming.

We overlook the issue that global warming is intertwined with many other air pollution problems. In addition to carbon dioxide, fossil fuels emit many toxic compounds into the atmosphere, with devastating impacts. Coal, for example, releases not only carbon dioxide but also a nasty brew of volatile organic compounds—nitrogen oxides, sulfur, mercury, arsenic, and radioactivity. Burning fossil fuels in motor vehicles causes smog, which hangs over large cities on hot summer days, exacting an enormous toll in deaths and health-care costs. Another blood relative of global warming is acid rain, caused by nitrogen and sulfur oxides, mostly from coal-fired power plants and metal smelters; these compounds damage lakes and degrade soil nutrients. In developed countries the battle against acid rain has gone well. In Asia, however, acid rain is running rampant, driven by the booming economies of China and India.

To save the atmosphere we must do far more than just seek "carbon sinks" and purchase "carbon offsets." Our goal must be to significantly reduce fossil-fuel emissions, and thus fossil-fuel use. But this is an enormous challenge since our growing economy is dependent on energy.

The degradation of the atmosphere is frightening. In spite of the Kyoto Protocol, greenhouse gases continue to increase and the temperature continues to rise. There is a disconnect between flicking on a light switch, a television, or an air conditioner and the consequences: air pollution, global warming, miners' deaths, and devastated mountain tops. Neither do we pay the price for those consequences when the electricity bill arrives each month. The situation is grim.

Yet I feel that it should not be so, for in the 1970s North America witnessed a renaissance in environmental awareness. Earth Day was born, regulatory agencies were formed, and environmental laws were passed. Measurable improvements followed. In spite of this, the fragile atmosphere continues to suffer. Clearly, far greater efforts are needed; but what are they? What path do we take?

Purpose and Content

The purpose of this book is to explain the complex topic of atmospheric pollution in a readable way. Another goal is to seek a path forward, to investigate whether and how we can save the atmosphere (and the planet).

The enormous complexity of the atmosphere and the processes occurring in it are explained. The cause and makeup of the four main atmospheric problems are described: the smog that hangs over cities, the acid rain that drifts over lakes and forests, the damage that is being inflicted on the protective ozone layer, and the most difficult of all, global warming and climate change. Other atmospheric problems such as Arctic and regional haze are also discussed. The scientific principles underlying these problems as well as illustrative case histories are presented. The history of these issues is also included for it is full of fascinating stories.

This book is written largely from a North American perspective although it also discusses issues from an international outlook, where appropriate. The *Système International* (SI) units are used throughout the book. Because the British system of units is so firmly entrenched in the United States, they are also presented for clarity (in brackets after the SI values).

I have come to the conclusion that meaningful progress requires not only the application of technological tools, but must also involve fundamental change in our lifestyles. We need to wrestle with seemingly sacrosanct issues: our ingrained suburban way of life, our dependence on the automobile, population growth, and unabated consumerism. We must utilize not only science and technology but must also wrestle with fundamental societal issues like seeking parity between rich and poor nations and, most importantly, realizing that continuing growth, the pillar of economic theory, has reached its limits and should be stopped. We cannot draw on the resources of the planet forever; we must seek equilibrium.

Unlike other books, *Air* places the important subject of air pollution into the broader societal context, and discusses policies and strategies involved in the air pollution field. I feel that we need to consider the science but also look at our own roles on the planet. I hope that by the end of this book, you will not only understand the issues, but will also want to take action and will feel confident in supporting steps to improve the situation.

Who It Is For

Air has been written in a non-mathematical style with numerous case histories, text boxes, and a glossary so it will be suitable for a wide readership. I hope it will also grace the desks of politicians, activists, and members of influential groups and will help them in their efforts to craft wise long-term policies.

Needless to say, saving the atmosphere will require dramatic changes and innovative thinking. We need to question established dogma, think outside the box, and seek fundamental solutions rather than just quick fixes.

1

The Plight of the Atmosphere

The conquest of the earth is not a pretty thing when you look into it too much.
—Joseph Conrad, *Heart of Darkness*

A few years ago my wife, Allyson, and I were hiking in the rolling hills that rise along the southeastern side of the immense island continent of Australia. Burnt dry and yellow by the relentless summer sun, the landscape was punctuated here and there by patches of eucalyptus trees in grays and pale greens. Keeping a watch for snakes, we climbed up a sun-bleached ridge. At the top we rested on a rock and drank in the view of sun-parched hills that rolled toward the horizon, seemingly forever. The visibility in the dry pure air was crystal clear and even distant birds stood out sharply and distinctly. Above us clouds formed crisp white mounds against a deep blue sky and the far horizon was awash in ribbons of purples and mauves.

The raucous screech of cockatoos echoed across the valley, and occasionally a small flock of parakeets burst from one tree and wheeled in formation to another, an incongruous flying patch of brilliant color against the monochrome landscape. Earlier we had seen a mob of kangaroos drinking cautiously at the edge of a dugout that collects rainwater for the livestock. A big gray—the lookout—sounded the alarm when we got too close and the mob bounced quickly down the valley like bewitched pogo sticks.

Almost all the trees around us were eucalyptuses, tall and stately guardians of this lonely landscape. Their sinuous trunks were largely exposed and naked, for the gum tree has the unusual habit of shedding its bark. Each trunk was a living art gallery, with the bark peeling and hanging like tattered medieval drapery, exposing patches of ghostly white wood covered with striking gray patterns and textures. Most bizarre were delicate reddish-brown hieroglyphics, the artwork of insects that had once burrowed below the bark.

We sat in silence, drinking from a canteen, savoring the moment. Allyson gazed for a long time, mesmerized by the landscape's stark beauty where nature is still untouched by the machinery of humans. She finally spoke: "When my time comes, I want some of my ashes to be sprinkled across these hills." I felt a similar powerful connection with the sweeping landscape.

After a while, I realized that a significant part of our enjoyment came from the simple act of breathing. The air on our ridge, which had traveled across thousands of kilometers of arid outback, unsullied by a single smokestack, was clean, fresh, and free of humidity. It was like inhaling an invigorating elixir, and we were lightheaded from this simple pleasure.

We take for granted—rarely notice, in fact—the never-ending act of filling and emptying our lungs. But that day I was very aware of breathing, of drawing in clean, pure air. I will always remember the smell of eucalyptus, the open spaces, the ghostly outlines of gum trees, and the intoxicating fresh air.

The Beauty and Complexity of the Atmosphere

The air we breathed in Australia is part of the gaseous fluid we call the atmosphere that surrounds the Earth. Although largely invisible, the atmosphere has its own raw beauty. One winter night while visiting Iqaluit in Canada's northern territory of Nunavut, I bundled up and climbed a snowy knoll. With my breath forming ghostly white puffs and the temperature hovering at −40°C (−40°F), I watched as the northern lights danced overhead in shimmering curtains of green. Although a dramatically different setting from Australia, I felt the same humility and wonder at the majesty of the atmosphere that surrounds our planet.

This atmosphere has a fascinating and complex personality. I am constantly surprised by its vitality, motion, amazing instability, and powerful mood changes. One moment it is a gentle breeze caressing your cheek, and that same afternoon the mood becomes gloomy as thunderclouds with their ominous anvil tops come rolling in. Soon after, lightning flashes across a darkened sky and thunder reverberates ominously down the valley. Sometimes powerful tornadoes or hurricanes can blast across communities, bringing surging waves and gusts with the potential to lift cars and destroy homes. Or the weather can quickly turn from a comfortable, pleasing temperature to paralyzing cold or stultifying heat. Air does not have the solidity and permanence of mountains and lakes; it is mobile and ever-changing.

Alfred Russel Wallace, co-founder with Charles Darwin of the theory of evolution, described the atmosphere as "the great aerial ocean"; this suggestive phrase captures the currents and eddies of the immense dynamic mass around and above us. Through constant motion, the atmosphere joins everything to everything else.

The atmosphere, which appears so ephemeral and delicate, is actually highly complex and performs a remarkable range of functions—some of which scientists have only recently begun to unravel. First and foremost, of course, it provides the reservoir and transport system for the vital carbon dioxide–oxygen cycle. Oxygen is the breath of life for humans and all other animals, from the smallest vole to the largest whale; and carbon dioxide, oxygen's companion gas, is essential for plants, from lichens to towering redwoods.

The atmosphere plays a key role in another great cycle. From the oceans and lakes it draws up that essential life ingredient, water, and—like a giant conveyor belt—distributes it again and again across the planet's land surface.

The atmosphere has built-in cleansing mechanisms able to remove air-borne pollutants created by humans as well as by nature itself. Rains wash contaminants out of the sky; trees and other plants cleanse the air and replace carbon dioxide with oxygen; "mixing" and turbulence ensure that natural toxic substances are diluted to tolerable levels. Over the eons, a multitude of systems have evolved, which work in a complex and symbiotic manner to preserve the biosphere—the Garden of Eden of all living things.

In addition to supporting life directly, the atmosphere acts as a protective shield. It shelters us from the fierce cold of outer space, for without the atmosphere the Earth would be a frozen ball of ice with an average temperature of –50°C (–58°F). It absorbs the deadly rays of sunlight while letting the beneficial ones pass through. It burns up millions of meteors each day before they can crash onto the planet's surface. And the ionosphere provides a layer against which radio waves bounce, enabling long distance communication.

It's hard to believe that the wispy, almost-invisible atmosphere can so profoundly affect our day-to-day life, given what a minute part of the planet it forms. The atmosphere is a very thin layer, with most of it (98 percent by mass) extending a mere 31 km (19 mi.) above the Earth's surface. In comparison, the Earth's radius (6,390 km/3,971 mi.) is over 200 times greater. Forming the tiniest one-millionth of the mass of the total global system, the atmosphere is a thin and tenuous layer. For this very reason it is also fragile and easily damaged.

Pollution

A few summers ago I looked back to the experiences my wife and I shared in Australia.

Niagara, where we lived, along with the rest of southern Ontario and western New York, was immersed in record-breaking heat. We trudged through an interminable series of oppressive days that dragged on with the temperatures rising above 30°C (86°F). A stifling humidity bore down mercilessly, leaving us listless and drenched with sweat. On those days we could barely distinguish blurry monochrome gray houses across the ravine behind our home, and across Lake Ontario smog hung over Toronto like a deadly brown dome. The air was so filled with pollutants that the health department issued advisories: the elderly and those with respiratory problems should remain indoors. Exercise and strenuous activity were to be avoided by everyone.

In contrast to Australia, breathing here was a tedious and difficult chore. I labored to pull the heavy air into my lungs and could almost feel the toxins drawn through the alveoli into my blood system; here they invaded corpuscles, accumulating in and damaging vital organs, and upsetting the delicate natural balance of this enormously complex system we call the human body.

Sadly, the bad air in Niagara is not an isolated case. Los Angeles is renowned for its traffic-induced blanket of smog, and virtually all megacities in the world suffer similar bad air. It is accepted by the inhabitants as an unavoidable cost of living in a large city. In many places, the pollution of the air is even worse. In the Czech Republic (formerly Czechoslovakia), the air was so bad the government paid a bonus to anyone who would live there for more than ten years; the locals called this "burial money" (Gore, 1992). In China, there are industrial cities so blighted by foul air the residents seldom see the sun. The national death toll due to outdoor air pollution is between 350,000 and 400,000 per year; and another 300,000 die each year from contaminated indoor air (Kahn and Yardley, 2007).

And we are helpless. We cannot stop breathing even when we are aware that millions of contaminants capable of causing life-shortening damage to our vulnerable bodies are floating in the air all around us. Air pollution is largely invisible; air is a gas in which we are completely immersed, so we have no choice but to draw it into our bodies every few seconds, every minute, every hour, every single day of our lives. If current trends continue, perhaps in the future we will live and work in air-tight homes and offices with sophisticated filter systems to cleanse the incoming air. This is not at all far-fetched; we are already taking similar actions with that other essential, life-giving commodity, water: think of the water filters and bottled water that have become so common.

The atmosphere has become seriously contaminated, especially in those places where there is significant human habitation. But wind carries pollutants even to those areas that are not urbanized. In Niagara, for example, many of the contaminants drift in on prevailing winds from coal-burning power plants in the US midwest. Acid rain from the industrial areas of Britain and Germany has ravaged lakes in Scandinavia; likewise, acid rain from the eastern United States travels to Bermuda. Windstorms in the Australian outback carry red dust onto the ice and snow of Mount Cook in New Zealand. Even the once-pristine Arctic has become fouled by industrial toxins. It is truly an age where we live on a connected planet—connected by air pollution as much as by the magic of technology.

More Than Just Contaminants

The litany of blows that humans are inflicting on the atmosphere is long. Not only is the atmosphere becoming increasingly loaded with substances that endanger human health; the very properties of the atmosphere are being altered. It is fast losing the ability to provide the conditions that make the Earth a comfortable place to live. The use of aerosol cans and refrigerants, for example, has caused the depletion of the ozone layer in the stratosphere, damaging the shield that protects us from the sun's deadly UV rays.

And now we have global warming, which presents far more of a challenge than any previous air-pollution problems. Modern society is built on a foundation of energy, most of which comes from burning fossil fuels. And when these fossil fuels are burned, they release carbon dioxide (CO_2) into the atmosphere, trapping the sun's heat and slowly raising the Earth's temperature. Global warming appears to be unavoidable given that the world is clamoring for more energy, and there is no technology currently available for capturing the carbon dioxide that emerges from smokestacks and tailpipes.

Even in Australia, where we were exhilarated by the freshness of the air, temperatures are inching upward—causing extreme drought and devastating bush fires. The range of temperatures to which the human body is acclimatized and for which we have designed homes and planted crops is changing. Can we adapt? Sea levels are rising and hurricanes are becoming more frequent. How do we cope?

These changes are taking place because we are throwing the atmosphere out of balance.

Air Pollution and Population

Disruption of the atmosphere is not new; it has been an inherent part of society ever since humans first learned to build fires and filled their caves with smoke. In the early days, fires were few and the atmosphere could dilute and cleanse the smoke that rose from them. Since then, humans have continued up the evolutionary ladder until we are now teetering on the top rung. With no natural predators, the population has exploded, aided by huge growth in technology. As the number of humans increases, the impact on the globe grows proportionally and, as one would expect, the impact is becoming more and more destructive. We have reached the point where the atmosphere is overwhelmed; no longer can it cleanse the "smoke" we are creating.

Our atmosphere, the canary in the mine, is wilting. The population has grown so large that it has exceeded the carrying capacity of the planet.

FIGURE 1-1

Another smoggy day in Los Angeles (photo by Ben Amstutz)

For the first time in history, human civilization is causing environmental changes at a global level, rather than just locally. Furthermore, with humans occupying almost all of the habitable areas of the planet, our ability to adapt and find solutions is limited. Solutions are difficult to implement, both technologically and politically. The atmosphere is sounding a warning that we must take action, rather than just reacting. But how should we proceed? What is the path?

THE CANARY IN THE COAL MINE

Early underground coal mines were primitive and dangerous places with no ventilation systems to remove lethal gases. Miners would bring a caged canary—a bird that is especially sensitive to gases such as methane (highly explosive) and carbon monoxide (highly poisonous)—into the workings to act as a warning system. The canary was carefully watched, and as long as it was well, the miners knew the air was safe. But when the canary became ill or died, the mine would be immediately evacuated. This early-warning system was used well into the twentieth century, and is credited with saving countless human lives.

Today's media are obsessed with global warming. It's the topic of the moment, and not a day passes without passionate reports of polar ice caps melting, heat waves scorching Europe, and island nations in the Pacific shrinking ever smaller to the bewilderment and anger of the inhabitants. The media, addicted to the 15-second news bite, debate whether a problem actually exists, and they sensationalize potential future scenarios.

Lost in this polemic, however, is the fact that global warming is only one symptom of the atmosphere's malaise. The overall degradation of the atmosphere, global warming included, is far more complex than the media can easily present. The atmosphere, for example, is not uniform. Rather, it is immensely complicated: air packets move in intricate patterns, sometimes rising as they warm, sometimes moving horizontally with the wind, other times forming a jet stream that bends back and forth across the mid-latitudes like a bullwhip cracked by a cowboy. And air is closely allied with the other great "lung" of the planet, the oceans. Water, thermal energy, and even chemicals move between the two great fluid bodies. At higher altitudes, the upper layers of the atmosphere are totally different from what we experience near the Earth's surface, consisting of ions and particles constantly buffeted by cosmic rays and influenced by the Earth's magnetic field.

To find solutions to global warming, we need to understand this complex atmosphere and how it behaves, a devilishly difficult undertaking. Furthermore, we must recognize that global warming is not an isolated issue. The media ignore the fact that it should be considered together with other atmospheric problems such as smog over cities, acid rain, and depletion of the stratospheric ozone layer—all these issues are interconnected like a giant spider web.

I need to sound another warning. Global warming is, in fact, just the latest in a series of increasingly serious atmospheric crises. As long as the human population continues to grow and technologies become more complex, it is inevitable that even more devastating, more difficult problems will arise in the future. Society seems to be incapable of dealing with global warming. How can we possibly deal with even more serious problems in the future?

Ultimately, the quest for solutions to the plight of the atmosphere is an investigation into the very nature of our civilization and its relationship to the global ecosystem. The deteriorating environment serves as a mirror in which we are able to see ourselves both as individuals and as a society. We need to ask some hard questions about our lifestyles and where our societies are heading. If we want our grandchildren to live in a world where doors are opening rather than closing, we will need to take action and make sacrifices. I hope we find the resolve.

But let us begin at the beginning and look at the atmosphere and how it functions. Without this knowledge, a search for solutions is meaningless.

2

The Delicate Atmosphere

Many little—and even not-so-little—boys and girls dream of being astronauts, of being strapped into the cockpit and feeling the awesome pull of G-forces as the rocket blazes into space. I know, for I have always wanted to go into orbit and see the Earth from a totally different perspective. I find photos from satellites captivating: an entire continent laid out like a map; the curve of a whitish moon pockmarked with craters; and the deep, deep blackness of outer space.

When astronauts return to Earth they talk of the breathtaking panoramas they have witnessed. And they often describe one striking feature: the soft glow that makes our planet look like a luminous blue teardrop. That blue glow is the atmosphere. Almost invisible from the ground, seen from space the atmosphere appears as an exquisite halo enclosing the planet.

What creates this halo effect? What is the structure and composition of the atmosphere? How did it form?

The History of Discovering the Atmosphere

People have always been fascinated with the possibility of flight. With the development of balloons, airplanes, rockets, and satellites, we became able to study the upper reaches of the atmosphere. In this way, the understanding of the atmosphere has grown hand in hand with the development of increasingly sophisticated instruments and our ability to carry them to greater altitudes.

The scientific understanding of the atmosphere began in 1593 when Galileo Galilei, tinkering with mercury inside a thin glass tube, invented the thermometer. In 1643, his student, Evangelista Torricelli, invented the barometer, an inverted column of mercury that enables atmospheric pressure to be measured. These were major breakthroughs; for the first time, these two key atmospheric properties could be measured quantitatively.

In 1648 Blaise Pascal carried a barometer up and down the stairs of the tower of Saint Jacques de la Boucherie in Paris, and was pleased to note that the column height changed (Sherman, 2004). Pascal, who was in poor health, persuaded his brother-in-law to carry the barometer up a nearby mountain and make careful observations (Young, 1977). As the brother-in-law ascended, he noted that the height of the mercury column decreased. These results were conclusive proof to Pascal that the atmos-. phere has mass and that its pressure decreases with height.

BLAISE PASCAL

French thinker, mathematician, and scientist Blaise Pascal (1623–1662) made major contributions not only in mathematics, but also in theology, philosophy, and atmospheric studies. A prodigy, Pascal mastered Euclid's *Elements* by the age of 12. In 1645, he invented the first calculating machine. His study of hydrostatics led to his inventions of the syringe and the hydraulic press, and greatly improved the understanding of the atmosphere. In 1647, because of his chronic poor health, he abandoned the study of mathematics and turned to gambling and a wanton lifestyle. Unprincipled behavior, perhaps; however, he used his gaming experience to contribute to the theory of probability! (*Britannica*, 2003).

During the latter half of the nineteenth century, hot-air balloons equipped with thermometers, barometers, and sampling bottles probed the upper levels of the atmosphere. (Kites were also used, but with much less success.) Those pioneering days were full of wild adventure. Balloons with frightened scientists clinging to the baskets were often buffeted by winds, caught in lightning storms, and immersed in ice as powerful updrafts shot them high into the air. Accidents, even deaths, were not uncommon. Around the globe these exploits captured the attention of the public, who followed every detail with morbid fascination.

It was soon discovered that the composition of the atmosphere is relatively constant to an altitude of about 10 km (6.2 mi.), consisting primarily of nitrogen, oxygen, and water vapor. Scientists subsequently discovered that the atmosphere also contains small amounts of solid materials such as grains of sand, bacteria, pollen, particles of soot, and salt crystals.

The ballooning explorations yielded a major discovery in 1899. A French meteorologist, Teisserenc de Bort, observed that at the very top of their ascents, his unmanned balloons measured an increase (rather than a continuing decrease) in temperature (Sherman, 2004). This went against the scientific consensus of the time, which postulated that temperature decreased with altitude at a rate of about 6 C° per kilometer (17 F° per mile), continuing to an altitude of approximately 43 km (27 mi.), where the temperature would reach −273°C (−459°F), or absolute zero. The only explanation was that a different, previously unobserved layer existed at higher altitudes.

De Bort proposed the term "troposphere" for the lower part of the atmosphere, from ground level up to about 15 km (9.3 mi.). The term is derived from the Greek word *trepein*, meaning "to turn over", as in the rising and falling of air currents. The newly discovered and mysterious region above the troposphere he designated the "stratosphere", from the Latin word *stratum* or "flat layer". He chose this name because this layer had no convection and, consequently, no mixing of air.

In the 1920s and 1930s the study of cosmic rays came into vogue. The Swiss physicist Auguste Piccard thought that the upper part of the troposphere shielded the Earth from most cosmic rays, which he believed were transformed to secondary rays at lower heights. To test his hypothesis, observations at high altitudes were needed. But due to a lack of oxygen, manned flights at such heights were not yet possible. Undaunted, he designed and built a balloon with an airtight aluminum gondola equipped with pressurized air (a feature that is of course now standard on airplanes).

In 1931, accompanied by his assistant, Piccard set off on an adventure whose progress was followed by people around the world. The pair overcame one problem after another, eventually rising to a record altitude of 15.8 km (9.8 mi.), where the atmospheric pressure is about one-tenth that at sea level. Having made history as the first human beings to enter the stratosphere, and having demonstrated that high-altitude flight is possible in pressurized cabins, they landed on a glacier high in the Tyrol and were rescued the following day (Sherman, 2004).

In 1932, in a redesigned cabin equipped with scientific apparatus and a radio, Piccard reached an altitude of 16.9 km (10.5 mi.). His instruments showed that cosmic rays were far more abundant in the stratosphere, which confirmed his hypothesis. He described the air at the height of his flight as "bluish purple, a deep violet … ten times darker than on Earth" (Sherman, 2004). In total, Piccard made 27 balloon flights contributing enormously to the understanding of the stratosphere (Gillispie, 1970). But the public was not impressed by the scientific aspect of his work. Instead, Piccard became famous merely for having risen higher than any other human, and was hailed as the "Columbus of the Stratosphere".

With the development in 1927 of the "radiosonde" (an inexpensive balloon-carried instrument package that radioed information on upper-level temperature, pressure, and humidity back to researchers on the ground), unmanned balloons could be sent aloft, allowing systematic mapping of the atmosphere's structure and circulation. By tracking the precise location of balloons with surveying instruments, researchers could also calculate wind speeds.

The highest altitude that can be reached by helium-filled balloons is the point where the density of the atmosphere decreases to the same density as helium. This occurs at a height of about 31 km (19.3 mi.). Higher altitudes can be reached only with rockets. The balloon's maximum was achieved in Project Man High I on August 20, 1957, by David Simons. He described how his balloon had been rising steadily for over two hours as if "on an endless elevator". Then the rate of ascent suddenly slowed and the balloon began to bob against the Earth's ceiling "like a basketball being dribbled in slow motion in an upside down world" (Young, 1990).

Before 1945, scientists studied the properties of the atmosphere above 31 km (19.3 mi.) using methods that were both ingenious and indirect. For example, by observing the altitudes at which meteors appeared and disappeared, they could determine high-level temperatures. The glowing sheets of northern lights provided the information that the atmosphere extended up more than 160 km (100 mi.). Temperature variations with altitude were determined by measuring the speed and behavior of sound waves caused by large explosions.

World War II brought two important new technologies. The first was the rocket, which allowed a giant leap in the altitudes at which data could be gathered. The second was radar, which has proven useful in tracking rain, snow, and important weather conditions such as hurricanes, tornadoes, and thunderstorms (Wolfe et al., 1966).

The rockets of the late 1940s could only ascend to a height just under 260 km (160 mi.), but the technology evolved rapidly. On October 4, 1957, only two months after David Simons had "dribbled" his balloon upside down, a rocket was launched from the Russian steppes that both shocked the world and transformed the study of the atmosphere. It carried the first satellite, Sputnik I—about the size of a basketball—and placed it into orbit about 1,600 km (994.2 mi.) above the Earth. It marked the start of the space age, and carried a payload of instruments to measure the Earth's magnetic field and the sun's energy and emissions.

By 1960, rockets were probing the entire atmosphere and satellites were regularly passing overhead. With the launching in 1960 of the first dedicated weather satellite, the Television and InfraRed Observation Satellite (TIROS I), scientists were able for the first time to obtain a truly global view of the atmosphere. TIROS orbited the Earth, transmitting cloud-

FIGURE 2-1

TIROS, launched in 1960, was America's first meteorological satellite (NASA)

cover pictures that enabled meteorologists to track, forecast, and analyze storms. With the success of TIROS, other weather satellites soon followed. The first geostationary weather satellite (GOES-1, with an orbit about 35,880 km (22,295 mi.) above the equator, allowing it to remain stationary over one point on Earth), was launched on October 16, 1975, and quickly became a critical part of the US National Weather Service.

Structure of the Atmosphere

Several centuries of scientific exploration of the atmosphere have provided an improved understanding of its structure. The atmosphere reaches several hundred kilometers into space. The bulk of its mass lies within about 30 km (18.6 mi.) of the Earth's surface, and its density decreases with height. Intuitively, one might expect the atmosphere to be relatively uniform with altitude (aside from diminishing density), but it is not. Instead, it forms four distinct layers, which are characterized by their temperatures. These layers—troposphere, stratosphere, mesosphere, and thermosphere—are separated by clearly defined boundaries.

The bottom layer, the troposphere, is the most important to humans (and life in general), given that it is the home of all living things. The temperature of the troposphere decreases by about 6 C° per kilometer (17 F° per mile) as altitude increases; this is because the troposphere derives its heat from the ground. Like water in a kettle on a stove, warm air near the ground rises, expands as it rises, and convects. Air movement in the troposphere is driven by vertical temperature differences as well as temperature variances between the equator and the poles. The height of the troposphere varies considerably from the equator (about 16 km/9.9 mi.) to the poles (about 10 km/6.2 mi.) and from season to season. The troposphere expands and contracts like a hot-air balloon; in the summer it is higher and in the winter it is lower.

The tropopause—the upper boundary of the troposphere—forms a kind of ceiling or barrier, and this does not allow most rising air currents to pass. This confines 85 percent of the troposphere below it. The flattened anvil shape you see on towering storm clouds marks the lower edge of the tropopause. The tropopause, however, is not a perfect seal and there are "leaky" regions where an interchange of air occurs between the troposphere and stratosphere.

The next layer is the stratosphere. Surprisingly, the temperature of the stratosphere increases with altitude, the opposite of what we find in the

FIGURE 2-2

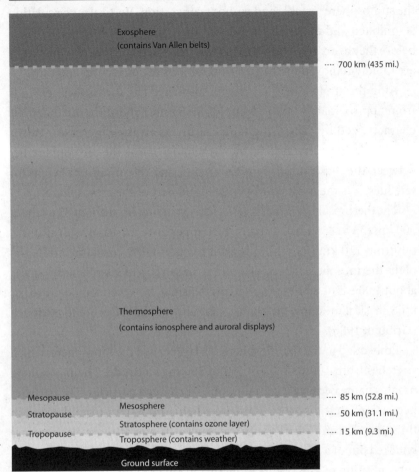

The layers of the atmosphere

troposphere. Scientists engaging in spectroscopy (the investigation of the spectrum of light) noticed in the nineteenth century that the shorter waves of ultraviolet sunlight were not as strong as they should be, and concluded that something in the atmosphere must be absorbing them. In 1913, Charles Fabry, a physics professor at the University of Marseilles, established that most atmospheric ozone lies in the stratosphere (Gillispie, 1970). This means that the temperature increase in the stratosphere is caused by ozone molecules, which absorb the sun's ultraviolet energy. Ozone also provides valuable protection, as ultraviolet light is damaging to living things.

Because the temperature increases with height and decreased pressure, the stratosphere is stable and without convection. Vertical air circulation is inhibited and materials introduced there may remain for many years before they are carried by gravity down into the troposphere (where they are rapidly removed by the cleansing actions of convection and rain).

Atop the stratosphere sits the mesosphere. The mesosphere extends from approximately 50 to 85 km (31 to 53 mi.) above the Earth, and is characterized by decreasing temperature as altitude increases (this is because there is very little ozone present). Because the mesosphere lies between the maximum height for aircraft and the minimum height for satellites, it is the most poorly understood part of the atmosphere.

The thermosphere is the largest layer, extending from about 85 to about 700 km (53 to 435 mi.). Initially the temperature increases with altitude, but from 150 km (93 mi.) upward it remains fairly constant. At the top of the thermosphere the temperature is the same as in interplanetary space, about 1,000°C (1,800°F). At such low densities, however, temperature has little physical meaning; for example, the air would not feel hot to the touch (Gribbin, 1986).

Embedded in the thermosphere is a layer called the ionosphere, which stretches from 100 to 400 km (60 to 250 mi.) altitude. The ionosphere has this name because it contains ions, or charged particles. The particles here are given a charge by ultraviolet light and X-rays from the sun; in this process, atmospheric gases are ionized as one or more electrons are knocked out of a molecule, leaving it with a net positive charge. The peak concentration of ions occurs at about 250 km (155 mi.), where about one in 2,000 particles is charged. Although this may not seem like a significant number, it is sufficient to reflect radio signals. There are three sublayers—one of which is sometimes observed to split—and they are named from the bottom up as the D, E, F1, and F2 regions. It is the F layer that in 1901 reflected Guglielmo Marconi's radio signals: they were bounced across the Atlantic, opening the door to international radio communications. (Today this technology has been largely replaced by satellite communications.)

The process of ionization is complicated by changes in the amount of solar radiation as the Earth revolves. The height and strength of the sublayers vary with the time of day and also with changes in emissions from the sun—for example, when solar flares occur. As the ions return to their neutral state, many of them give off light. Although this glow is too faint to be seen from Earth, satellites orbiting in the darkness of space have taken photographs of the ionosphere's glow.

SIR EDWARD APPLETON AND RADIO COMMUNICATION

Appleton was born in 1892 and educated at the University of Cambridge. He worked at the Cavendish Laboratory, Cambridge, from 1920 until 1924 and then went to King's College, University of London, where he gained an international reputation for his research into the ionosphere and the propagation of electromagnetic waves. Appleton showed that radio waves with a sufficiently short wavelength could penetrate the lower part of the ionosphere and would be reflected off a higher layer, which became known as the F Region, or Appleton Layer. His discovery made long-range radio communication possible and assisted in the development of radar.

The aurora borealis (northern lights) and aurora australis (southern lights) are beautiful, dynamic displays seen in the night sky—one of the greatest natural shows on Earth. The most common form of an aurora is a green curtain-like luminosity extending east to west. Aurorae occur in the thermosphere and generally extend from about 100 km (62 mi.) in altitude to about 300 km (186 mi.) or higher. Seen from above, the auroral curtains appear as ovals surrounding the geomagnetic poles, but with the center displaced slightly toward the dark side of the Earth. These ovals, with an average radius of about 2,200 km (1,370 mi.), extend thousands of kilometers east to west but are only about 1 km (0.6 mi.) thick from north to south.

The aurorae are particularly spectacular in winter and many Japanese make pilgrimages to Alaska and northern Canada to witness the shimmering sheets of light dancing across a star-filled sky. Japanese people also visit because they believe that a child conceived under the northern lights will be especially intelligent.

Investigations have revealed the aurorae are strongest at times of sunspot activity, when the sun ejects charged particles—the solar wind—that travel outward and interact with the Earth's magnetic field. These energetic particles (protons, electrons, and ions) are channeled toward the poles by the magnetic field; they collide with and ionize atoms and molecules in the thermosphere. This causes the auroral luminosity. The most common green color is emitted by energized oxygen atoms. The occasional crimson color results from excited nitrogen molecules.

SPIRIT TORCHES

Rather than looking for explanations in magnetic fields and ions, the Inuit of the north believe that the luminous colors of the aurora represent torches borne by spirits who are lighting the way to heaven for those who have died violently. The Inuit see woven into the auroral display spirits playing a game of football with a walrus skull. The sound they hear is that of the spirits' footsteps crackling on the snow.

The exosphere, which lies beyond the thermosphere, above 700 km (435 mi.), is an austere region with very sparse gases. An atom must travel several kilometers before bumping into its nearest neighbor. It is notable primarily for the magnetic field, which is caused by the rotation of the molten nickel-iron core in the center of the Earth. The magnetic field extends throughout the planet and well into surrounding space. The volume influenced by the Earth's magnetic field is called the "magnetosphere". At a distance of about 60,000 km (37,000 mi.) from the Earth toward the sun, the pressure of the solar wind is balanced by the geomagnetic field. This serves as an obstacle to the solar wind, which is largely deflected around the Earth. The magnetosphere streams out into an elongated "magnetotail" stretching several million kilometers from the Earth away from the sun.

High in the exosphere are the Van Allen belts. These are two doughnut-shaped zones centered on the equator, which trap and hold appreciable numbers of energetic protons and electrons in the magnetic field. The inner belt extends from roughly 1,000 to 5,000 km (620 to 3,100 mi.) above the terrestrial surface and the outer belt from some 15,000 to 25,000 km (9,300 to 15,500 mi.). The belts are named after James A. Van Allen, the American physicist who discovered them by accident in 1958. He detected the particles with Geiger counters aboard the earliest US scientific satellites, Explorer 1, Explorer 4, and Pioneer 3, which measured the flux of cosmic rays in space. It was the first great discovery of the space age.

The protons in the inner Van Allen belt have such high energy that they can penetrate several centimeters of lead. They are thought to originate from the decay of neutrons produced by the interaction of the lower atmosphere with high-energy cosmic rays coming from outside our solar system. Some of the neutrons are ejected upward, and a fraction of them transform into protons and electrons in the Van Allen belts, where they are trapped by the Earth's magnetic field. Collisions with atoms in the thin atmosphere eventually remove the particles, but they generally survive for about ten years; this allows them to accumulate and form the radiation belts (*Britannica*, 2003).

VAN ALLEN BELTS A GALACTIC HOAX?

Conspiracy theorists believe that because of their lethal levels of radiation, the Van Allen belts form an impenetrable barrier to humans. Thus, they reason—and I use this word loosely—that the landing of humans on the moon, the International Space Station, and similar achievements are enormous hoaxes perpetrated by the FBI and the CIA so the United States can control the world.

Composition of the Atmosphere

The lowest part of the atmosphere, the troposphere, is composed primarily of nitrogen (78 percent) and oxygen (21 percent). Argon, an inert noble gas, accounts for about 0.9 percent. The remaining 0.1 percent is composed of many trace gases, most significantly carbon dioxide (CO_2) and water vapor. Table 2-1 shows the concentrations of the components of the troposphere. Obviously many of the physical and chemical processes that occur in the atmosphere are directly related to its composition.

The concentrations of the three main gases (nitrogen, oxygen, and argon) are relatively constant up to about 100 km (62 mi.). Although carbon dioxide makes up only 360 parts per million (ppm) of the atmosphere by volume, it is vital in maintaining the atmospheric heat balance given its strong absorption of infrared (thermal) radiation.

Another important trace gas is ozone (O_3), a molecule composed of three oxygen atoms. Ozone has a split personality. In contrast to oxygen (O_2), which is essential to life, ozone is poisonous. It is found in minute quantities near the Earth's surface, where it is an ingredient of smog. In the higher atmosphere, however, ozone forms a shield that protects life on Earth

TABLE 2-1: Composition of the Atmosphere at Ground Level
(after Emiliani, 1992)

Compound	Concentration (% by volume)
Nitrogen	78.1
Oxygen	20.9
Argon	0.93
Carbon dioxide	0.036
Neon	0.00182
Helium	0.000524
Methane	0.000155
Krypton	0.000114
Hydrogen	0.000055
Xenon	0.0000087

from UV rays. Its greatest concentration—only about 12 ppm—is centered in a layer at about 25 km (16 mi.) altitude. To give some idea of how thin and fragile this ozone layer is, if compressed to liquid and placed at sea level it would have a thickness less than 0.5 cm (0.2 in.) (Schaefer and Day, 1981). Nevertheless, this tenuous ozone layer plays a vital role in biological activity, given its absorption of deadly UV radiation.

As those who live in humid areas know, air is never perfectly dry. Water vapor is usually present in highly variable quantities ranging from a fraction of 1 percent by volume in dry desert regions to more than 4 percent in humid tropical regions. Survivors of floods and tropical inundations caused by rain would find it hard to believe that if all the water in the atmosphere were condensed and spread evenly over the Earth, it would make a puddle only about 3 cm (1.2 in.) deep (Young, 1977). Compare this to the oceans, which if spread evenly over a flat Earth, would make a layer 2.7 km (1.7 mi.) deep. Only one one-thousandth of one percent of the water on Earth resides in the atmosphere. Yet this tiny amount plays a critical role in that complex phenomenon we call weather.

Water is a simple yet remarkable substance. Its importance cannot be overstated, for without it there would be no atmosphere nor any life as we know it. Although made from only two hydrogen atoms and one oxygen atom (compared to a molecule of table sugar with 45 atoms, or to a simple protein molecule with two or three thousand atoms) it is one of the most important substances on the planet. Useful properties of water include its ability to dissolve and carry other substances. Another important property is its high heat capacity compared to other geologic materials: in other words, it takes more heat to raise the temperature of water than that of soil or rock. Water vapor plays an important role in maintaining life. It absorbs considerable infrared radiation and is an essential link in the hydrologic cycle. Through cloud formation and precipitation, water is a fundamental component of weather.

In addition to gases and water vapor, air contains particles of solid matter. Airborne particles are formed by several processes (Schaefer and Day, 1981). Larger pieces of material, such as rocks, are broken down by mechanical and chemical weathering until they can be lifted and carried by the wind. These particles are generally 5 microns (a micron is a millionth of a meter) and larger because smaller pieces will adhere to each other to form larger ones. Another process begins in a vapor phase, forming particles through condensation and crystallization. These particles are usually less than one micron in size. Soot particles can also form from combustion. Generally, only those particles that stay in the air for more than an hour are considered to be important (usually smaller than about 50 microns).

The atmosphere is—to a degree—self-cleansing. Once the concentration of particles exceeds a certain limit (about 100,000 particles per cubic centimeter or 1.6 million particles per cubic inch) the smaller ones coagulate through random motion until they are so large they drop to the ground. In cloud formations, particles larger than 0.1 microns act as "condensation nuclei", around which raindrops or snow crystals form. Many of these particles are picked up by falling rain or snow.

Evolution of the Atmosphere

The development of the atmosphere is intimately tied to the evolution of the planet and of life itself. Air is the breath of the planet, emitted by volcanoes, as well as the respiration of living plants, animals, and even tiny bacteria that reside deep in rock, soil, or the sea.

The Earth was probably formed from the collision of numerous small objects called "planetesimals" and meteorites, which were without their own atmosphere. Scientists generally agree that the atmosphere was formed later by gases escaping from rocks, usually caused by localized heating, melting, and volcanic action. Here, however, a problem arises. The composition of volcanic gases is different from the composition of today's atmosphere. In particular, volcanoes emit virtually no oxygen. Today's volcanoes spew out primarily water (up to 97 percent) and carbon dioxide with lesser amounts of nitrogen, hydrogen, sulfur dioxide, and chlorine. There are also smaller amounts of hydrogen sulfide, carbon monoxide, methane, ammonia, and other gases. There is no evidence to suggest this composition was different in the early history of the Earth.

Much of the light hydrogen would have escaped the gravitational attraction of the Earth. Although considerable amounts of carbon dioxide have been emitted by volcanoes over the eons, very little (only about 0.03 percent) is in the atmosphere today, with the rest primarily "stored" in carbonate and shale rocks, the oceans, and plants (Gribbin, 1986).

Volcanic emissions, therefore, can account for the present composition of the atmosphere, except for its oxygen. There is a good deal of evidence that oxygen was completely absent during the first billion years of the Earth's existence. Then where did it come from?

One method by which oxygen is produced is through the breakdown of water vapor by ultraviolet light, called "photodissociation", which yields free hydrogen and oxygen. If this had occurred, however, large quantities of hydrogen would have to escape Earth's gravity. Scientists consider this scenario to be unlikely.

The more likely source of oxygen is life itself, mainly through photosynthesis—the process in which plants combine carbon dioxide and water to produce carbohydrates and oxygen. Studies have shown that the early atmospheric composition, when stimulated by lightning or sunlight, could yield the complex organic compounds such as amino acids that are necessary for life to evolve (Gribbin, 1986). According to some estimates, approximately 99 percent of the total amount of free oxygen added to the atmosphere since the Earth's beginnings was produced by photosynthesis, with only 1 percent produced by photodissociation.

Liquid water first appeared on the planet about 3.8 billion years ago (about 0.7 billion years after the Earth was formed) and calculations show that the water vapor released by volcanic activity since that time is more than enough to account for the oceans. The emergence of water in the liquid state was probably the single most important factor in the evolution of the planet and life itself. It affected not only living organisms by providing a habitat, but also the temperature of the planet, by influencing solar reflection and absorption through the oceans, ice, and clouds. Water has been an important factor in keeping global temperature in the 7°C–27°C (45°F–81°F) range throughout Earth's history, even though the sun's heat output has increased about 30 percent over this period (Gribbin, 1986).

About 3.5 billion years ago, in the shallow but growing oceans, life forms began to take shape and reproduce. For over two billion years, bacteria were the only known life form. Photosynthesis began producing free oxygen about 3.5 billion years ago with the evolution of cyanobacteria, and has continued producing it in increasing amounts to the present day. At the beginning of the Cambrian era (570 million years ago), the concentration of oxygen in the atmosphere was about 2 percent. Today it is about 21 percent.

Earth's climate has changed over time. The earliest ice age is believed to have occurred around 2.7–2.3 billion years ago. Since then there have been recurring periods of glaciation, particularly in the northern hemisphere. The most recent peaked about 18,000 years ago. Ice ages are thought to be caused by small changes in solar radiation, the result of changes in the orientation of the Earth as it orbits the sun. In the more distant past the Earth was too warm to support such large-scale ice formation (Environment Canada, 1990). The warming at the end of the most recent ice age has continued for several thousand years. This is a very abrupt change on the geological time scale, although of course in terms of human perception it seems very gradual.

In the early days of the planet, living organisms were protected from deadly UV radiation by a blanket of ocean water. With the gradual accumulation of oxygen in the air, an ozone layer began to form in the stratosphere, providing more protection to the incipient life forms in the oceans. When the ozone layer had developed to near-present levels some 500 million years ago, life was able to leave the protective embrace of the oceans to populate the land. Over the next 300 million years, the surface of the Earth gradually turned green as mosses, ferns, pine and spruce forests, and giant redwood and sequoia trees proliferated.

About 100 million years ago an important new class of plants arrived: angiosperms, the seed-bearing plants. The seed of an angiosperm is a complete embryonic plant, which can travel far from its parent. In addition, plants had developed a new method of reproduction that did not rely on the wind, but instead used attractive colors and fragrances to lure insects, bees, and birds to pollinate them. This led to grasses, flowers, and blooming shrubs and trees. But most importantly, the angiosperms yielded fruits and grains, which provided food for the warm-blooded animals that were beginning to appear on Earth.

This process of evolution continued its long journey until it reached the stage at which the species *Homo sapiens* appeared. As humans evolved, they developed ever more powerful tools until in recent centuries the vast and dirty engines of our civilization have begun to significantly alter the exquisite blue halo observed by astronauts.

3
The Ever-Changing Weather

One day, as I worked in my office, thick flakes of snow swirled outside the window. Visibility had closed in and only the nearest fir trees could be seen, with their branches drooping under coats of white. The panorama I had become accustomed to—islands and forests sparkling in sunshine—had suddenly been transformed. I was pleasantly surprised. I missed the white winters of eastern Canada and here in the mild climate of the Gulf Islands on the west coast of British Columbia, snow is rarely seen. The weather forecast the previous evening had predicted light rain followed by sunny skies. I could only smile and enjoy the still whiteness— the reality is that the vicissitudes of weather are beyond our control.

Not only is weather fascinating with its ever-changing and often unpredictable moods but, together with the climate, it has an enormous impact on humans and our activities. It affects the crops we grow, the types of houses we live in, the clothes we wear, our outdoor activities, the sports we enjoy, and much of the conversation around the office.

Weather is defined as the state of the atmosphere at any given time. Climate is the average weather over a long period of time. The energy that creates weather and climate comes from just one source: the sun—that enormous nuclear power plant around which the planets revolve. As the Earth's tilt, rotation, and orbit around the sun vary, so does the amount of solar energy received by our planet. Without the Earth's tilt at an angle of 23.5°, the sun would always shine directly over the equator and there would be no seasons. Instead, there is more sunshine north of the equator for half of the year, and more south of the equator for the other half.

The atmosphere, along with the oceans, plays a critical role in how solar energy is received and distributed around the globe. First, it acts like the glass in a greenhouse, letting in and then trapping solar energy, which heats the ground and oceans before slowly trickling back into outer space. Without the atmosphere, the Earth would be as cold and barren as the moon.

Global Winds

The second important contribution of the atmosphere is to disperse and mix the energy received from the sun. Imagine our world with no wind: it would be a harsh and forbidding place. Intense temperature extremes would exist between the equator and the poles, and between night and day. No rain showers would sweep over land, and inland regions would be barren rockscapes untouched by soil-building processes.

The main winds on a global scale are caused primarily by temperature variances between the equator and the poles. In fact, the polar-equatorial temperature difference is considered one of the pillars of climate. If no other effects were involved, a simple circulation of hot air rising at the equator and sinking at the poles would be established. But there is a major complication to this model: the spinning of the Earth. The globe's rotation causes air moving from high to low pressure in the northern hemisphere to deflect to the right. In the southern hemisphere air moving from high to low pressure is deflected to the left. This is called the "Coriolis force", and gives weather systems their rotation—for example, making hurricanes spin like tops. The amount of deflection is related to both wind speed and latitude. Gentle winds will be deflected only slightly while stronger winds will be deflected more. Similarly, winds near the poles will be deflected more than winds of the same speed closer to the equator. (At the equator, the Coriolis force is zero.)

What causes the Coriolis force? Imagine a giant cannon shooting a cannonball directly north from the equator. As it travels north, the cannonball will curve to the right. To understand this, consider that on the equator the eastward speed of the Earth's rotation is greatest, at Washington less, at Montreal even less, until at the North Pole there is no eastward motion at all. When the cannonball is shot it has not only the northward component of speed imparted by the cannon, but also an eastward component caused by the spinning of the Earth. As the cannonball flies northward, it is moving toward the east with progressively more speed than the ground below, so it curves to the right. The Coriolis force acts on the wind in a similar manner, causing curving and circular flows.

The polar-equatorial thermal difference combined with the Coriolis effect sets up large-scale atmospheric motions that drive regional wind patterns. The main global winds are caused by air masses in the equatorial regions rising because of solar heating, creating a low-pressure zone

near the ground. As these rising air masses reach the top of the troposphere, they flow toward the poles. But due to the Coriolis force, they are
also deflected eastward: by the time they reach 25° to 35° latitude, the air
at high altitude is traveling almost due east. As the air cools and sinks it
creates a high-pressure zone, pushing the air in the lower atmosphere out
of the way. Some of the surface air flows south toward the low-pressure
area near the equator to replace the air that has risen. But this near-sur-
face air is deflected westwards as it travels, creating the trade winds. This
wind pattern, so well known by sailors, assisted Christopher Columbus
in his discovery of North America.

GUSTAVE-GASPARD CORIOLIS: BILLIARD-PLAYING MATHEMATICIAN

Born in 1792 in Paris, Coriolis became an engineer and mathematician and was the first to describe the effect of motion on a rotating
body, a concept of enormous significance in meteorology, ballistics,
and oceanography. He worked at the École Polytechnique in Paris
from 1816 to 1838 where he coined the terms "mechanical work" and
"kinetic energy". In 1835, he published his masterpiece, in which he
showed that on a rotating surface there is inertial force acting on the
body at right angles to the direction of motion—this is in addition to
the ordinary effects of motion of a body. This Coriolis force affects
wind direction, as well as the rotation of hurricanes and tornadoes.
That Coriolis enjoyed the occasional break from his scientific work is
demonstrated by his treatise, *Mathematical Theory of the Game of
Billiards* (*Britannica*, 2003).

At the equator there is a belt called the "Intertropical Convergence
Zone", where the northern and southern trade winds come together. This
zone, also known as the "equatorial trough" or "doldrums", has little wind
and was much feared by sailors in the days of sailing by wind power alone.
Without the hint of a breeze, mariners could be trapped for long periods
in the stultifying heat and humidity.

The latitudes around 25° to 35°, where the downward air flow dominates, have almost continual high-pressure conditions. These regions are
characterized by fair weather, clear skies, and little rain, given that most
of the moisture was condensed out of the atmosphere when the air rose

to high altitudes near the equator. These are known as the "horse lati-tudes": it is said that sailors would be becalmed so long in them that they were forced to eat the horses on board their ships. The world's most for-midable deserts—the Arabian, Sahara, Kalahari, Gobi, and Great Australian deserts—lie along these two zones.

Because air is primarily moving downward in these zones, it retains fumes and air pollution for long periods. Los Angeles, at 34° latitude, is a city blessed with good climate but it also suffers from intense pollution, primarily due to cars.

As air descends in the band of 25° to 35° latitude, not all of it travels toward the equator. A fraction of the subtropical air moves toward the poles. The bands between 35° and 60° latitude are known as the temper-ate zones, or mid-latitudes. In the early days of climatology it was thought that as a result of the Coriolis force, their air masses curved eastward as they moved toward the polar regions. This matched the observations near the ground—for example, that much of the northern United States and most of Canada have prevailing southwesterly winds, a phenomenon that helped ships sail from North America to Europe.

In the polar regions, cold air descends, causing high pressure that forces surface winds down and outward to the mid-latitudes, where the polar air meets the warm, humid subtropical air. A broad band of turbulence is thus created in each hemisphere, with warm and cold fronts moving on prevailing westerly currents around the world.

In this way, the northern and southern hemispheres each have three major circulation patterns: the Hadley cell (a circular pattern of air motion), the Rossby waves, and the Polar cell (see Figure 3-1). The dif-ferences in heat received from the sun at the equator and the poles are averaged, moderating the extreme temperatures that would otherwise occur. In the tropics, the Hadley cell winds, near ground level, are pre-dominantly from the east (the trade winds). In the mid-latitudes the winds near the ground are primarily from the west. Near the poles, winds are generally from the east, forming part of the Polar cell. Using this infor-mation, early climatologists could construct a general pattern of global weather that reflected an elegant symmetry.

Of course, life is not as simple as Figure 3-1 would indicate. Neither the Intertropical Convergence Zone nor the boundaries between the global cells are constant; instead their width and location vary as they

FIGURE 3-1

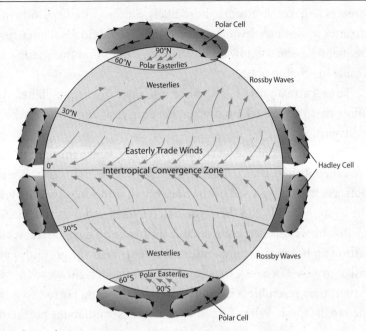

Global wind circulation patterns

are influenced by the movements of large air masses and the changes in seasons. In particular, detailed observations in the mid-latitudes revealed that a cell was not present. Instead, winds in this zone flow westward at all heights, even at high altitudes where they had been predicted to flow eastward. The winds are centered on a core, or jet, of high-speed wind located near the top of the troposphere that changes position with time.

Scientists confirmed the presence of this "jet stream" in the 1920s and 1930s using weather balloons. During World War II, bomber pilots encountered these jet streams and made more detailed observations. After the war, studies revealed that the jet streams form at mid-latitudes at heights of approximately 10 km (6.2 mi.) with average speeds of 60 km/hour (37 mph) in summer and 150 km/hour (93 mph) in winter. Speeds of up to 400 km/hour (250 mph) have been recorded. The jet streams are about 2–4 km (1.2–2.5 mi.) thick vertically, about 50–150 km (31–93 mi.) wide horizontally, and are located at approximately 50° latitude in summer. They move closer to the equator in the winter.

Photographs from satellites have helped in the understanding of jet streams. Cirrus clouds are more likely on the equatorial side of the jet streams, while the poleward sides are often free of clouds. The cirrus clouds frequently cast a shadow on lower-lying clouds and this shadow is easily visible in satellite photos.

The jet stream also has a major influence on airline flights. Airplane pilots try to piggyback on it when heading east, and avoid it when flying westward.

The westerlies of the jet streams have a wavelike motion; these waves are called Rossby waves. Instead of heat moving poleward via a convection cell, it is transported by the horizontal, sinuous, wavelike motions of the westerly winds and their associated eddies and disturbances.

Rossby waves exert considerable influence on day-to-day weather, by setting up horizontal temperature gradients near the ground that create polar fronts. The westerlies rarely flow in a straight west-to-east line; rather, they resemble a meandering river, looping far to the north and then to the south. When a Rossby wave snakes south in the northern hemisphere, it can bring polar fronts and cold weather as far south as the southern United States. This north-south movement also plays a role in forming storms that carry strong winds and large amounts of snow or rain.

BIZARRE JET-STREAM ATTACK

During World War II the jet stream played a major role in one of the most diabolically clever wartime attacks ever mounted. Soon after Japanese high-altitude bombers discovered the jet stream in 1944, their military devised an unusual plan. They sent 9,000 balloons aloft into the jet stream that traveled eastwards across the Pacific Ocean to mainland United States. The unmanned balloons, armed with bombs that would detonate automatically when the balloons descended, were equipped with a mechanism that kept them in the jet stream at an altitude between 9 and 11 km (5.6 and 6.8 mi.). Due to one small oversight—the electronic equipment was powered by a wet-cell battery, which froze in the high altitudes—only a handful of balloons actually landed and detonated. There were six fatalities on May 5, 1945: five children and their teacher on a Sunday school outing to the seaside near Bly, Oregon (Lynch, 2002). One can only imagine the amazement and puzzlement of the local inhabitants, not to mention the US military command, since this bombardment apparently materialized out of nowhere. Only several years after hostilities ceased was the mystery of the balloon attack unraveled (Young, 1977).

During the Cold War the US military copied the Japanese tactic and used the jet stream to carry unmanned balloons with spy cameras over the Soviet Union (Lynch, 2002).

Weather

Weather affects us in countless ways. It influences whether we stay inside or go outdoors, whether we carry an umbrella, when we plant and reap, and even whether we are happy or melancholy.

Understanding weather is closely related to communications, for information has to be compiled from many weather stations spread over large geographic areas. In 1870, President Ulysses Grant approved a congressional resolution to found a US national weather service. Soon 24 weather stations were operating, managed by the US Army, with information communicated via telegraph. To inform people of incoming weather conditions, signal flags were flown from flagpoles on public buildings (Kidd and Kidd, 1998).

Progress since then has moved at an amazing speed. Developments in remote-sensing satellites, communication methods, and high-speed computers have made it possible to study the atmosphere, construct mathematical models of complex processes, and forecast the weather with increasing sophistication and accuracy. Today these capabilities allow seven-day weather forecasts, including: tracking major hurricanes, alerting orange and grape growers of impending frost, providing advance warning of blizzards, and advising sunbathers and beachgoers about high levels of UV rays.

A network of weather stations covers the entire world. Every day at the same time around the world, at 12:00 and 24:00 Greenwich Mean Time, approximately 700 radiosonde balloons are launched with miniature weather stations, which record and transmit information as they rise. All this data—together with measurements from dozens of satellites, about 10,000 ground stations, 7,000 merchant ships, 300 moored and drifting buoys, and hundreds of aircraft—are fed into computers that perform incredibly complex calculations to model the weather. We may be skeptical of the weather forecasts we see each evening on television, but they are soundly based on principles of science and detailed observations of the atmosphere around the globe.

Let's look at some of the key components of weather.

Air Pressure

Atmospheric pressure (the weight of the atmosphere on a given unit area) is the single most important factor to consider when predicting what kind of weather the immediate future holds. Pressure is measured using a barometer, normally calibrated in units of one one-thousandth of a bar, or a millibar (mb). The average atmospheric pressure at the surface of the Earth is 1,013.2 mb, and pressure values range from less than 940 mb in very intense lows to over 1,050 mb in very strong highs (Wolfe et al., 1966).

THE BAR AND MILLIBAR

The bar is a unit of pressure that is equal to 100 kilopascals, where a pascal is a force of one newton per square meter. "Bar" is derived from the Greek word *baros*, meaning "weight". The bar and millibar (mb, one one-thousandth of a bar) were first introduced by Sir Napier Shaw in 1909. Because one bar is approximately the pressure of the atmosphere at sea level and changes in atmospheric pressure are conveniently expressed in millibars, the bar and mb are commonly used by meteorologists and atmospheric scientists.

In addition to measuring pressure, it is important to consider how it changes. Rapidly rising pressure brings the best weather; cool upper air sinks, and as it descends and compresses, becomes warmer and the ice crystals and raindrops that form clouds evaporate, usually resulting in sunny skies. In contrast, rapidly falling pressure brings the worst weather; warm moist air rises and soon cools to saturation producing huge towers of clouds covering large areas with prolonged rain and dark skies.

The upcurrents of a low-pressure system, which are much stronger than the downdraughts of a high-pressure system, often bring strong winds and storms. To maintain a global balance between the amounts of rising and falling air masses, high-pressure systems (known as "anti-cyclones" or highs) cover much larger areas of the Earth than low-pressure systems (known as "cyclones", depressions, or lows). Because of their extensive size, highs are less mobile than lows. That's why stormy weather can be here today and gone tomorrow, whereas the settled weather of high pressure zones tends to last longer.

Winds

Until it became possible to measure wind speed and direction, scales for indicating wind strength were rudimentary. The invention in 1846 of the "cup anemometer", or wind gauge, greatly expanded meteorologists' ability to understand wind (de Villiers, 2006). Scientists found that winds generally follow the patterns established by the global circulation patterns, but vary significantly depending on local conditions and geography.

Local winds, for example, can be caused by the daily cycle of temperature differences between land and water. During the day, the land warms up more than the sea or a large lake does, because rocks require less heat to raise their temperatures than does water. For this reason, air over land rises, and air from over the water flows in to replace it; this results in a daytime sea or lake breeze. To compensate, air at higher levels moves toward the water. The reverse happens at night, causing surface winds to flow toward the water.

Just as in the daily cycle of winds along a coast, regional winds can be caused by seasonal temperature differences between land masses and seas. This is the cause of the monsoons in India: during the winter, winds flow toward the ocean, and during the summer they flow toward land.

MONSOON PARTY

In India, the arrival of the monsoon is an important event because it brings the rains that turn the land green and nourish crops. The prediction of the monsoon's arrival is anxiously followed, and as the onset draws near, the meteorological center at Trivandrum, on the southern tip of India, becomes the site of a media frenzy. A party atmosphere prevails when people clutching umbrellas line the shore and await the dark clouds that will bring four months of rain.

Local conditions influence the character of wind. "Catabatic winds" become warmer and drier as they descend a mountainside and the atmospheric pressure increases; the increasing temperature lowers relative humidity. Examples are the *foehn* that blows down the southern side of the Alps and the Santa Ana that blows down the southwest slope of the coastal ranges of southern California. The "chinook", or "snow eater", that blows down the eastern slope of the Rocky Mountains can bring—especially in winter—astonishing changes. On January 27, 1962, for example, the temperature at Pincher Creek, Alberta, rose from –19°C (–2°F) to +3°C (37°F) in less than an hour. With the wind gusting to 95 km/hour (59 mph) the snow cover disappeared as if by magic.

DOES THE WEATHER INFLUENCE YOUR HEALTH?

Biometeorology, a specialized branch of science combining biology and weather research, is trying to answer that question. When I lived in Calgary, Alberta, we loved the arrival of a chinook wind and the temporary reprieve it brought to the frigid winter. But it was a mixed blessing; I, and many others, often developed mild headaches. Our symptoms have been verified by studies that showed migraine sufferers have headaches during and just before chinooks.

Associated with chinook and Santa Ana winds is an electrical disturbance called "atmospherics", electromagnetic waves with frequencies in the 4,000–50,000 cycles per second range. Recent research has demonstrated a distinct correlation between low-frequency atmospherics and electrical activity in the brain of a headache sufferer. These atmospherics also affect memory, mental ability, and attention span (Lynch, 2002).

WINDBREAKS

Studies have shown that wind passing through 30 m (98.4 ft) of mature forest loses 20–40 percent of its force, and wind passing through 100 m (328 ft) of forest loses about 90 percent of its force. Trees are thus efficient windbreaks, as farmers have known for centuries.

To track weather, meteorologists plot the distribution of pressure on maps known as "isobaric charts". Pressures are shown as contours, or isobars, often spaced at 4 mb intervals, which form circles or ellipses around low- and high-pressure systems. Air tries to flow directly from a high-pressure area to a low-pressure area; but because of the Coriolis force the motion is altered, so that in the northern hemisphere the winds circulate clockwise around a high and counter-clockwise around a low. So instead of wind flowing perpendicular to isobars, it flows almost parallel to them.

WHERE ARE THE HIGHS AND LOWS?

At the middle latitudes of the northern hemisphere, if you place your back to the wind, the low-pressure area is to your left. This is known as the Buys-Ballot Law.

Air masses—large volumes of air that move in a body and have relatively uniform temperature and pressure—play an important role in determining the weather. They are classified according to origin as "polar", "arctic", "tropical", or "equatorial" (from coldest to warmest). The humidity of an air mass is determined by whether it has traveled over land or the sea. The former is named "continental" (dry) and the latter "maritime" (humid). A continental polar air mass is cold and dry, and a maritime tropical air mass is humid and warm. Continental polar and arctic air masses often sweep from northern Canada through the central and eastern United States, bringing cold air as far south as the Gulf states. In the south, maritime tropical air masses often travel up the Mississippi valley in the summer reaching as far north as Canada.

The boundaries between air masses are called "fronts"; these are areas of contrasting temperatures, high "wind shear" (when winds of different directions come together), and active weather. A "cold front" occurs when a cold air mass (say, a polar air mass) is advancing and displacing a warm air mass (for example, a tropical air mass). As the front moves forward, the warm air is forced to rise over the denser cold air, forming a boundary with a slope of about 1:100 (that is, a vertical rise of 1 meter every 100 horizontal meters, or 1 yard every 100 yards). If the warm air is moist, convective clouds and thunderstorms can be produced. Cold and warm fronts are shown in Figure 3-2.

When the winds cause a warm air mass to move toward a cold air mass, a "warm front" occurs. The slope of a warm front, about 1:200, is shallower than that of a cold front. The warm air slides slowly up over the

FIGURE 3-2

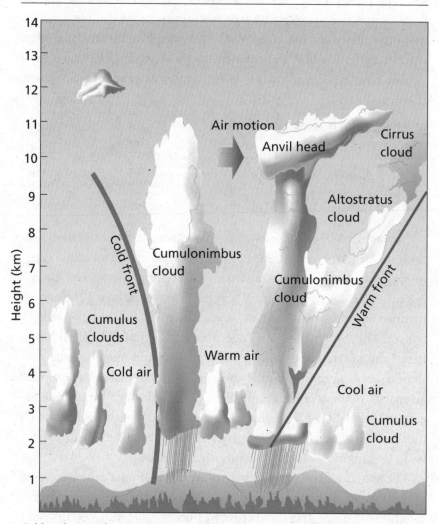

Cold and warm fronts with associated cloud formations (Wolfe et al., 1966)

cold air and produces widespread cloud layers; this results in long-lasting rain. The front may be 15–30 km (9–19 mi.) wide and stretch for thousands of kilometers.

A relatively common occurrence is for a small irregularity to form along a front. This often grows to become a wave composed of a warm front and a cold front with a low-pressure zone centered on the spot where they meet. Sometimes the cold front overtakes and pushes under the warm front, and the portion of the cold front extending beyond the curve of the wave is called an "occluded front". When the weather pattern has reached this stage it is called a fully developed "cyclone". This term (sometimes also called a "depression") denotes an atmospheric pressure system around which winds rotate (in the northern hemisphere, the rotation is counter-clockwise).

FATHER OF METEOROLOGY

The credit for developing many of the fundamental principles of meteorology rests with Norwegian physicist Vilhelm Bjerknes, who did his formative work between 1917 and 1922. Working at a geophysical institute in Bergen, Norway, with a team of scientists, he discovered the basic mechanisms of the rainstorm. He recognized that air masses operate on a global scale and, in addition, he gained an understanding of cold and warm fronts (Sherman, 2004).

Clouds

Henry David Thoreau wrote, "A sky without clouds is a meadow without flowers, a sea without sails." Astronauts looking down on the Earth have been impressed by the unexpectedly large amount of cloud, covering about half the globe, and moving restlessly across the continents and oceans.

Clouds are composed of water drops. As we saw earlier, dust particles and other foreign matter act as nuclei on which water droplets grow. Salt attracts water, and so forms a particularly good nucleus. Other kinds of nuclei—such as those derived from acids from forest fires and coal-fired power stations—also attract water. A cloud forms when the concentration of water droplets reaches about 1,000 per cubic centimeter (about 16,400 per cubic inch). The droplets are so small that turbulence keeps them suspended in the air; they continue to grow until they become too large to remain in suspension.

The four main cloud types, as first proposed in 1803 by the English naturalist and pharmacist Luke Howard, are based on their appearance from the ground (Bryson, 2003). Their names are derived from Latin: *cumulus*, meaning "heap" or "pile"; *stratus*, meaning "flat"; *cirrus*, meaning "curl"; and *nimbus*, meaning "rain". These four terms, used in various combinations and modified by words such as *alto* (meaning "raised" or "elevated"), describe the ten different types of cloud that are agreed upon internationally. These ten cloud types are also key indicators for predicting the approaching weather; additionally, they are divided into "high", "middle", and "low" levels.

Cirrus clouds only occur at high altitudes from 8 to 15 km (5 to 9 mi.) and are white, as they consist of very small ice crystals. They appear as fine, white filaments that are wispy or feathery (for this reason they are sometimes called "mares' tails"). The trail of ice crystals may appear to originate from a distinctly denser head, or may be stretched into long streamers several kilometers long. When cirrus clouds form in long bands they may indicate the direction of high-speed jet streams. Due to their height, cirrus clouds catch and reflect the sun's rays immediately before sunrise and after sunset, while the rest of the world is still in darkness. The appearance of cirrus clouds is often an early warning of an approaching weather front.

"Cirrostratus" is a sheet of cloud that may cover large parts of the sky, but the sheet is so thin that the sun can be clearly seen through it. "Cirrocumulus"—a rarer phenomenon—are small, round puffs that are often aligned in long rippling rows. They create beautiful sunsets as they catch the light of the setting sun (Lynch, 2002).

In the mid-level sky, where the temperature is warmer, clouds are formed mainly by raindrops. "Altocumulus" clouds are white or gray fluffy masses that spread out in waves across the sky. Their appearance on a summer morning often signals thunderclouds by the afternoon. "Altostratus" clouds are vast gray sheets. When they drop lower and cover a large area, prolonged rain may follow (Lynch, 2002).

Stratus clouds form a flat layer that fills the sky. They are created when a warm air mass passes over a colder air mass, and they generally bring damp drizzle rather than rain. If they are several kilometers thick with rain they are called "nimbostratus", and the sky will be dark and gloomy. "Stratocumulus" clouds look sparse and lumpy, and they range from light to dark gray. Patches of blue sky sometimes peek through. This cloud type is often seen in late afternoon when larger cumulus clouds break up. Stratus clouds do not have any significant electrical activity.

Rising hot humid air, or "thermals", yield cumulus clouds, normally a sign of fair weather. These individual clouds, formed by convection, look like piles of white froth. Their sharply defined bottoms mark the altitude where the condensation temperature occurs. Although these clouds may appear large and imposing when seen from the ground, a cumulus cloud of several hundred meters diameter is actually quite insubstantial. In fact, it contains only about enough water to fill a bathtub (Bryson, 2003).

HOW DO RAINDROPS FORM?

This question baffled meteorologists for many years, because raindrops are more than a million times bigger than cloud droplets. We have seen how cloud droplets form as water gathers around an infinitesimally small nucleus. But laboratory studies and theoretical calculations showed that cloud droplets would not naturally coalesce into raindrops. The explanation is that as temperatures drop below 0°C (32°F), rather than freezing, the droplets in a cloud remain liquid in a supercooled state. Occasionally, one water droplet freezes and then very rapidly it attracts other droplets that also freeze. As the single ice crystal grows into a snowflake, it gains weight and falls through the cloud, colliding with cloud droplets and leaving a trail of ice pieces to start the formation of yet more snowflakes. In this way, a few ice particles can quickly grow into a snowstorm inside a cloud. As the flakes fall into warmer temperatures, they melt and become rain (Young, 1990).

Because of the avalanche effect of ice crystal formation, some scientists think that clouds could be seeded with particles that have a structure similar to that of ice—such as silver iodide—and would thus be "tricked" into producing rain. Although many experiments have been tried, artificial rain-making has not yet evolved to a commercially viable level.

SILENT AS A SNOWFLAKE?

We can all picture a peaceful winter scene with deep drifts of snow covering the landscape and snowflakes falling gently and silently from the sky. But if you could stand near a body of water—perhaps a pond that hasn't yet frozen, or a fast-running stream—and tune your ears to a higher frequency than humans normally perceive, you would hear a cacophony of tiny screams. The structure of a snowflake is such that it traps a significant amount of air. On landing in water, the ice melts and a bubble of air is released. Before the bubble can rise to the surface, the surface tension and water pressure cause the bubble to vibrate at a frequency far higher than the human ear can detect. Each dying snowflake thus emits a tiny scream lasting only one ten-thousandth of a second.

"Cumulonimbus" or thunder clouds form from cumulus clouds during good weather as the sun heats the ground surface, increasing the convective rate. As the sun beats down, the cumulus clouds grow larger and tower higher. The lower surface of a cloud may be only 1 km (0.6 mi.) from the ground, but its top may burgeon up as high as 9–12 km (5.6–7.5 mi.), into the region of the jet stream, which shears off the top of the cloud. This gives it the distinctive flat-topped anvil shape. Serene as it may appear from the outside, the inside of the cloud is a turbulent seething mass of currents, powerful updrafts, and eddies (Farrand, 1990).

Collisions between water droplets and ice crystals cause the cloud to become electrically charged. The reasons are not well understood, but lighter particles tend to gain positive charges and are carried aloft, while heavier particles become negatively charged and accumulate near the bottom of the cloud (Bryson, 2003). The potential difference between top and bottom can be an astonishing 100 million volts (compared to about 150,000 volts for a high-voltage transmission line). The once docile cumulus cloud has now become a violent and dangerous cumulonimbus cloud. Forked lightning begins to dart within the cloud and from the base of the cloud to the ground. First, a relatively weak leader stroke travels from cloud to ground, and then a massive stroke returns to the cloud along the ionized path. A bolt of lightning can travel at 435,000 km/hour (270,000 mph) while heating up the air immediately around it to 28,000°C (50,400°F)—good reasons to get off the golf course as soon as possible when thunderheads approach! Heavy rain pours from the base of the cloud. As the thunderstorm reaches maturity, the updrafts are replaced by downdrafts, which cut off the supply of heat and moisture that has fed the rain.

Soon the storm weakens. The thunder and lightning become irregular and die away, and the heavy rain is replaced by light showers. Eventually the cumulonimbus, having exhausted itself, drifts away and is broken up by the wind. Cumulonimbus clouds often form in late afternoon following the

TABLE 3-1: Summary of Cloud Types

Level	Name	Description	Coming Weather
High	Cirrus	White, wispy	Weather front
	Cirrostratus	White, thin sheet	Rain or snow when thickening
	Cirrocumulus	White, puffy, often in rows of ripples	
Mid	Altocumulus	White or gray fluffy mass	Possible thunderclouds in afternoon
	Altostratus	Large, even gray sheet	Prolonged rain when decreasing in height
Low	Stratus	Even, flat gray sheet	Dampness and drizzle
	Nimbostratus	Solid gray	Continuous light rain or snow
	Stratocumulus	Lumpy gray with patches of blue sky	
	Cumulus	Fluffy cotton wool	Fair weather, but can develop into nimbocumulus in afternoon
	Cumulonimbus	Towering, gray, puffy	Thunderstorms, rain

rising of thermals. They also occur along advancing cold fronts. Approximately 1,800 thunderstorms are taking place at any given time around the world. About 400 people are killed in the United States each year by lightning, and about seven in Canada (Environment Canada, 2004a).

Two unusual cloud types, "nacreous", meaning "mother of pearl", from the French word *nacre*, and "noctilucent", meaning "shining in the night", from the Latin *nocti* (night) and the Latin *lucere* (to shine), occur in the stratosphere at very high altitudes of 20–30 km (12–19 mi.) and 80–90 km (50–56 mi.), respectively. They are formed by ice crystals. Due to the great height of noctilucent clouds, it is believed that their ice particles form around debris left by meteors and meteorites (Young, 1990).

URBAN HEAT ISLANDS—HUMANS INFLUENCE WEATHER

In the early 1970s, the city of St. Louis was selected for a meteorological experiment because it is surrounded by flat land with no nearby large water bodies. Results from the study showed that rainfall in an area about 15–50 km (9–31 mi.) east of the city (in the direction of the prevailing winds) was 20 percent higher than in the surrounding area. In addition, the temperature in St. Louis was about 1.6–3 C° (2.9–5.4 F°) higher than in the surrounding areas. The researchers concluded that large cities produce their own weather.

Compared with nearby countryside, cities have lighter winds, less humidity, less sunshine, and more cloud, rain, and thunderstorms. The most significant influence, however, is on temperature: factories, cars, buildings, and people give off large quantities of heat. This heat, as well as energy from the sun, is absorbed by brick, asphalt, and concrete during the day and radiated back into the air at night. The effect is most marked during clear cold nights and winter mornings. In Montreal, the city is sometimes 5–8 C° (9–14.4 F°) warmer than its surroundings. In summer, daytime temperatures are 1–2 C° (1.8–2.6 F°) higher in urban areas.

The urban hot zone causes a rising column of air, which creates clouds and—since the city air is heavily laden with pollution that act as nuclei—increases rain. Because the city-made clouds tend to rise higher than those created in the surrounding area, they generate increased and more violent thunderstorms with more rainfall and more hail (Young 1977, 1990).

In northern climates the urban warmth brings some blessings. Toronto, for example, has approximately 15 percent lower heating bills, less snowfall, and a longer gardening season than the immediate neighboring areas (Phillips, 1990).

Hurricanes

Air has a "Jekyll-and-Hyde" personality. Because it is invisible, we think of air—if we think about it at all—as benign and delicately wispy. But we should never forget that this benign air can sometimes transform into a monster. The atmosphere has substantial mass (about 8.8 million tonnes for every square kilometer [about 21 million tons per square mile] of the planet) and when that bulk is set in motion enormous forces can be created. There is no better example than hurricanes, or typhoons as they are called in the Pacific Ocean, which are very large cyclones. They might be considered thunderstorms multiplied many times in strength and having a spiral spin. It is hard to believe that so tenuous a thing as air can have such enormous power, with torrential rains and winds that can exceed 300 km/hour (186 mph).

A hurricane forms when tropical air is unstable and the Intertropical Convergence Zone is displaced over unusually warm waters (surface temperature greater than 27°C/81°F and warm to a depth of 70 m/230 ft or more) and is also far enough away from the equator for the Coriolis force to take effect (above about 5° latitude). The Atlantic Ocean just off Africa, where hot air from sub-Saharan Africa meets cooler ocean air, is a notable breeding ground for hurricanes. With so much thermal energy stored in the sea, immense amounts of water are evaporated into the air—as much as 2 billion tonnes (2.2 billion tons) of water a day—and are carried upwards in a spiral of strong winds (Young, 1977). The process reinforces itself as the enormous amount of heat released by the vapor condensing into clouds makes the air even more buoyant. Towering cumulonimbus clouds form, relieving pressure at the water's surface, causing further air convergence with intense spiraling winds. In this way a hurricane can grow to about 150–750 km (93–466 mi.) in diameter. Hurricanes are on average about 550 km (340 mi.) across; the "eye", in the middle of the hurricane, is about 20–50 km (12–31 mi.).

FIRST (AND SECOND) FLIGHT INTO THE EYE

In 1943, US Air Force pilot Joseph Duckworth became the first person to fly a plane into the heart of a hurricane and record the awesome range of conditions between the calm eye and the powerful, turbulent outer winds. On returning to base, he described the extraordinary flight to the station's weather officer, who became eager to experience it himself. So Duckworth flew through the hurricane a second time with the weather officer on board (Lynch, 2002).

Tornadoes, or twisters, form under the same conditions as severe thunderstorms and squall lines. High humidity, extreme instability, and marked surface air convergence are all factors. Tornadoes often occur in regions where cold fronts and warm fronts collide, such as the Great Plains of North America, where polar air from Canada often meets tropical air masses coming north from the Gulf of Mexico. In fact, the central United States has about 700 tornadoes per year, more than anywhere else in the world. Tornadoes contain such enormous power they are able to lift freight cars and strip trees of their leaves and bark. They often occur in "swarms". For example, in early April 1974, 148 tornadoes were reported in just over two days stretching from Ontario to Alabama. The inside of a tornado funnel is marked by very low pressure; this creates an intense suction, causing much of a tornado's devastation (Lynch, 2002).

THE BEST AND WORST CLIMATE IN THE UNITED STATES

The Farmers' Almanac used sunshine, sky conditions, precipitation, humidity, and wind to determine the "worst" and "best" cities for weather. Astoria, Oregon, is the worst; it is the cloudiest in the nation (240 days) and is third in terms of wetness (177 cm/69.6 in. per year). Tied for second-worst are Marquette and Sault St. Marie, both in Michigan. They rank among the coldest and snowiest US cities and are also among the rainiest (Farmers' Almanac, 2006a).

For the best weather head to Yuma, Arizona—number one because average precipitation is a mere 6.7 cm (2.65 in.) per year, rain only falls 17 days per year, and it is sunny 90 percent of daylight hours. It is among the least humid cities, with an average relative humidity of just 38 percent. However, summer temperatures average over 38°C (100°F) from June 4 to September 24. But it's a dry heat.

Tied for second place for the best weather are Las Vegas, Nevada, and Phoenix, Arizona (Farmers' Almanac, 2006b).

THE BEST AND WORST CLIMATE IN CANADA

If you are looking for bracing and character-building weather head for St. John's, Newfoundland, the city with the worst weather. Of major Canadian cities, St. John's is the foggiest (124 days), snowiest (359 cm per year/141 in. per year), wettest (151 cm per year/59 in. per year), windiest (24.3 km/hour/15.1 mph average), and cloudiest (only 1,497 hours of sunshine per year). Furthermore, storms frequently blast the city (Phillips, 1990).

The balmiest weather in the country is found on Vancouver Island and the nearby Gulf Islands in British Columbia. Based on 23 "comfort" weather categories such as mildest winter, the most sunshine, the lightest winds, and the lowest humidity, Victoria and Nanaimo tied for first place when 100 Canadian cities were ranked (Environment Canada, 2004c).

Carbon Dioxide–Oxygen Cycle

The atmosphere is a vital part of life and controls two essential processes: the carbon dioxide–oxygen (CO_2–O_2) cycle, and the water cycle (or "hydrologic cycle").

The CO_2–O_2 cycle is essential for all living things, whether plant or animal. There are two fundamental stages. In the first, plants, using chlorophyll as a catalyst, take carbon dioxide (CO_2) from the atmosphere, water (H_2O) from the ground, and energy from the sun to form glucose ($C_6H_{12}O_6$) and oxygen (O_2). The oxygen is released to the atmosphere while the glucose is stored in the plant. This process is shown in equation form as:

$$6CO_2 + 6H_2O + \text{Solar Energy} \longrightarrow C_6H_{12}O_6 + 6O_2$$

Glucose in turn is converted by plants into starches, fats, and other large molecules that store energy. The main oxygen production is by green plants such as forests on land (about 35 percent) and phytoplankton in the ocean (about 65 percent).

The cycle is completed when animals, including humans, breathe oxygen and eat plants. In digestion, animals break down the large molecules such as glucose to release stored energy and form the simpler molecules CO_2 and H_2O. The carbon dioxide is released to the atmosphere for the plants to reuse. In some cases, plants die and are covered by soil and rock, to be resurrected in the distant future as coal and oil.

Over the eons, a balance has been reached on the Earth where the amount of carbon dioxide produced by animals and the amount of oxygen produced by plants appears to be in equilibrium. In this cycle each part depends on the other for survival. And the atmosphere plays a key role in the process; it is the reservoir containing these two life-giving gases.

Humans, however, are throwing a wrench into this cycle. The atmosphere is becoming polluted with a range of toxic compounds, which disrupt the delicate processes. Of still greater concern is the fact that we are burning the huge reservoirs of coal, oil, and natural gas that formed over hundreds of millions of years, releasing vast quantities of carbon dioxide into the atmosphere. The equilibrium that has existed for a large fraction of the planet's lifespan is now, suddenly and jarringly, being thrown out of balance.

The Water Cycle

Water is one of the most essential materials on the Earth for many reasons. First, it governs the climate: without the large quantities of water on the planet, temperatures would be far more variable. Second, water is a principal agent in the geologic processes of erosion, transportation, and deposition. It sculpts valleys and forms new rock layers that later become mountain ranges. Third, water is essential for life, for without it plants and animals could not survive. Indeed, the human body is 60 percent water—salty, in fact, like sea water. Fourth, water is a key factor in human activities. It is used for transportation, irrigation, recreation, and the creation of energy.

In the water cycle, clouds are the essential bucket brigade. Water enters the atmosphere by evaporation, mostly from the oceans, but also from lakes, rivers, and living things. Clouds carry the moisture over the oceans onto the land, where it falls as rain and snow and then eventually flows back to the ocean, completing the cycle. Water moves continuously, circling from the Earth to the atmosphere and back to the Earth.

Table 3-2 shows that oceans are the largest reservoir in the water cycle. Indeed, oceans are the powerhouse that drives the globe's surface behavior. Meteorologists are increasingly treating the atmosphere and oceans as a unified—but highly complex—system. The water cycle is dynamic and in constant motion. Each year almost 1 meter (1 yard) of water is evaporated from the surface of the world's oceans. Most of this water returns by direct precipitation onto oceans, but about 7 percent falls on land and is returned by runoff from rivers. Because of the large surface area of the oceans, this prodigious amount of water is constantly rising and falling.

TABLE 3-2: Depths of Water Layers on a Smooth Earth (in meters)

Atmospheric water	0.03 (1.2 in.)
Lakes and rivers	1 (1.1 yards)
Ground water	10 (11 yards)
Land ice	45 (49 yards)
Oceans	2,685 (1.67 mi.)

Although the amount of water contained in the atmosphere forms only 0.001 percent of the total water on the planet, this tiny amount is critical to the operation of the overall water cycle. Without atmospheric water, the rest of the water cycle would not take place.

The atmosphere is fascinating, complex, and full of mysterious and wondrous delights ranging from delicate clouds to northern lights to thunderous typhoons. Let us now look at how our human activities are affecting this thin gaseous layer upon which all life depends.

4

Smog: The Deadly Haze

I don't like smog, for it has significantly affected my personal life. The Niagara area of southern Ontario is famous not only for the thundering cataracts of Niagara Falls but also for its thick, spirit-sapping humidity, mixed with a hefty dose of ozone, particulate matter (discussed in more detail on pages 61–63), and sulfur dioxide—the lethal ingredients of smog. The year 2002 was a pivotal one as we sweltered in a particularly nasty outbreak of heat and humidity. Over the summer the Niagara Health Department issued no less than 21 smog advisories: warnings that asthmatics and others with breathing problems should stay indoors and that everyone should avoid outdoor exercise. In addition, 16 days had ground-level ozone concentrations exceeding 50 parts per billion (ppb) (the benchmark for an advisory), and another eight advisories were issued for excessive heat (Reid, 2002). For those like me who enjoy hiking and jogging, it was an unbearably long and unpleasant summer. For those with asthma or lung conditions, it was the summer from hell. One stultifying afternoon, my wife and I reached a difficult decision: we would leave family and friends behind and seek a better climate. In 2005 we moved to the west coast of Canada.

Smog is impossible to ignore. Its brownish haze is clearly visible and it makes breathing difficult. Especially on sunny days it forms an unpleasant fog that from afar looks like an enormous drab mist hovering over the city. Under calm conditions, this dark polluted air can extend upward for 1.5 km (0.9 mi.). Under windy conditions the dome transforms into a plume that can stretch downwind for more than 30 km (19 mi.).

I've always been uncomfortable with the comparison of smog to an umbrella. Smog does not offer protection from rain or sleet. Instead, sinister chemicals lurk in that brownish haze, preying on the elderly, asthmatics, and those with other respiratory problems. The heavy humid air makes us listless and can affect the breathing process. Also—make no mistake—smog is deadly, extracting a heavy toll in illnesses and fatalities.

As with other forms of air pollution, smog is created by the machinery of society, with the main culprit being transportation. Particularly worrisome is the fact that there is no obvious cure for foul air; or more appropriately, there are remedies but we refuse the medicine. Cities, which have become "pollution islands", continue to grow larger and larger. Our factories continue to manufacture more and more motor vehicles as well as trucks, trains, and planes.

Historical Cases

Smog has evolved and changed with the times. When smog first appeared at the peak of the Industrial Revolution in the late 1800s, it was mostly sulfurous. In that era, the smoke that poured from industrial smokestacks and household coal fires contained significant amounts of sulfur dioxide (SO_2). The sulfur dioxide combined with water in the fog to form sulfuric acid. The acid in turn joined with fine particles of soot to form a serious health hazard. The term "smog", a combination of "smoke" and "fog", was coined in 1905 by Dr. Henry Antoine Des Voeux, a British public health official. Six years later he made another significant observation when he noted that smog was deadly and that it had claimed many lives in Edinburgh and Glasgow.

Episodes of smog were common in England in the late nineteenth century and were nicknamed "peasoupers". The most famous episode of air pollution was not that long ago. Starting on December 3, 1952, a temperature inversion (a phenomenon in which warm air stays near the ground instead of rising) coupled with still winds held a dome of toxic pollutants—caused primarily by the smoke from burning coal—over the city of London for about a week. When the killer smog finally lifted, it left behind over 4,000 dead. Such "peasoupers" are unlikely to arise again since British legislation has banned coal fires in homes and required industries to build higher smokestacks. However, the experience underlines the importance of protecting air quality. It is also a testimony to the danger of burning coal.

There have been many other episodes of deadly air pollution. In 1930, in the industrial Meuse Valley of Belgium, 64 deaths were recorded due to severe smog (Environment Canada, 1990). Fatal cases of smog are not restricted to Europe. In 1948 in Donora, Pennsylvania, a heavy smog was created when releases of sulfur dioxide, carbon monoxide, and metal dust from local steel plants and zinc works were held near the ground by a temperature inversion. It was so smoggy that houses across the street were not visible. Nearly 7,000 people, about half the town's population, became ill, and about 50 people died (Tollefson et al., 2000). In 1966, New York City suffered a severe smog incident causing 128 deaths.

These cases show that since the beginning of industrialization, smog has exacted a severe human toll. And it continues to do so.

California has long suffered poor quality air as a result of motor cars. In 1947 the governor of California signed the Air Pollution Control Act. This was the first such law in the United States; it mandated the establishment of air pollution control districts in every county of California (Goodell, 2006). In 1969 California again led the way by passing air quality standards for key pollutants including sulfur dioxide, nitrogen dioxide, and particulates.

The year 1970 was a banner year for the environment. Riding the emotional tide of millions of supporters, Earth Day was created. Sensing the national mood, President Richard Nixon formed the US Environmental Protection Agency in the same year (Goodell, 2006). These were important steps, and protecting the environment is an ongoing battle. Sadly, the battle against smog often seems to be a lost cause.

Causes of Smog

The character of smog changed as coal was gradually replaced by the cleaner fossil fuels, natural gas and petroleum. Although sulfurous smog is still common and quite dangerous in some developing countries, a new kind of smog emerged in the richer countries after World War II. It is almost wholly linked to the enormous growth in the use of automobiles.

Smog became a problem in California as early as the mid-1940s. On July 26, 1943, for example, in the midst of a heat wave, a pall of smoke and fumes descended on downtown Los Angeles, reducing visibility and gripping people with a nearly unbearable eye-stinging, throat-scraping sensation (AQMD, 1997). Mistakenly, local industry was blamed. In 1948, Professor Arie Haagen-Smit, a chemistry professor at the California Institute of Technology in Pasadena, made a major breakthrough. For some time, farmers had complained that air pollution was damaging their crops, bleaching or discoloring the leaves of plants—something not seen in other parts of the country. Haagen-Smit examined the plants and found that the smog in southern California was not a mixture of smoke and fog (as the name had always implied); but instead it was composed largely of ground-level ozone, a reactive and damaging chemical. To everyone's surprise, the primary cause of California's air pollution was revealed to be not industry, but cars.

This relatively new form of air pollution, called "photochemical smog", is created when three conditions occur together. The first is heavy automobile traffic, which emits several nitrogen oxides (denoted NOx) and volatile organic compounds (VOCs; primarily unburned gasoline). The second, surprisingly, is bright sunlight and warmth—not what is normally associated with pollution. This is why the chemical reactions that result in smog are called "photochemical reactions" (from the Greek, *photo*, meaning "light"). These two conditions happen relatively frequently in cities, which as we have seen act as heat islands; and, of course, cities also contain vast amounts of traffic.

The third condition is a stagnant air mass, which is caused by meteorological conditions and landforms. For example, cities located in bowls or valleys such as Mexico City and Los Angeles are more likely to suffer from incidences of severe smog, and this is especially the case when an atmospheric temperature inversion acts as a lid preventing dispersal of the pollution. As we saw in Chapter 3, the situation in Los Angeles is aggravated by its location in the mid-latitudes (the areas lying between approximately 30° and 60° north or south of the equator) where prevailing high pressure yields long periods of calm with little wind.

NOx are highly reactive gases that form ozone—toxic to plants and animals—in the presence of sunlight. In addition, they form nitric acid, nitrate salts, and toxic organic nitrates. The photochemical reactions also produce small quantities of other chemicals that can harm living things and the environment. Photochemical reactions are described in greater detail later on in this chapter.

Particulate matter, consisting of fine dust, soot, and aerosols, is another key element of smog. Sulfur dioxide, a pollutant which is always present to some extent, oxidizes and hydrates to form sulfates and sulfuric acid that become part of the particulate matter in the air. Automobiles also emit much particulate material into the air, as well as carbon monoxide (CO), one of smog's most toxic constituents.

Instead of being thanked for making this important discovery, Haagen-Smit was vilified for associating car exhaust and California's legendary sunshine with the pervasive smog. He was attacked by both auto manufacturers and car lovers. After considerable debate, an independent commission in 1957 vindicated Haagen-Smit's conclusions (Sherman, 2004).

In the 1950s, the smog situation in Los Angeles was becoming intolerable, and it led to a longstanding battle to improve air quality that continues today. Along the way numerous control measures have been implemented, including banning backyard incinerators, eliminating orchard smudge pots (which prevent overnight freezing of fruit), installing vapor recovery systems on gasoline station fuel-pump hoses, equipping cars to reburn crankcase blowby gases (the cause of one-third of vehicle emissions), developing cleaner gasolines, and introducing catalytic converters on all cars.

In spite of these substantial improvements, the metropolitan Los Angeles area—the nation's leader in developing stringent air-quality regulations—still has the worst air in the United States. Formidable progress has been made but it has all been swamped by an ever-growing population driving an ever-growing number of vehicles.

And the situation is even worse in Mexico City, one of the world's largest cities. Here emission controls on cars were not required until 1992. In addition, the combustion of gasoline at Mexico City's elevation of 2,240 m (7,350 ft) is only two-thirds as efficient as it is at sea level. Not surprisingly, during the winter (when warm air passing over the mountains gets trapped in the valley and stagnates) air pollution increases considerably, endangering life and health.

Fossil-fuel power plants, especially those that burn coal, are also major emitters of smog-causing pollutants and contribute as well to acid rain and global warming. For example, it is estimated that about 50 percent of the smog in southern Ontario, including my old stomping ground of Niagara, comes from the industrial midwest of the United States with its numerous industries and coal-fired power plants. (We look more at these power plants in Chapter 7.)

Smog has grown to become an international problem. Virtually every major city in the world suffers, especially during the warmer seasons, including Los Angeles, Mexico City, Toronto, Phoenix, Rio de Janeiro, London, and—as we saw during the 2008 Olympics—Beijing.

Health and Environmental Impacts
Fortunately, deadly episodes of smog such as the famous London "pea-souper" of 1952 no longer happen in developed countries. Smog, however,

continues to be a major health hazard and takes a surprisingly large death toll—but does so quietly. Through history, in fact, smog has adversely affected the health of more people than any other type of pollution.

Photochemical smog causes eye irritation, lacrimation (excessive secretion of tears), raw throat, and lung inflammation in humans, and severe damage to many types of vegetation, including important crops. Air pollution also damages buildings and works of art such as sculptures. But smog has a deadlier side. It increases mortality—especially among those who suffer from respiratory and coronary ailments. It also contributes to many types of cancers.

First, the good news. The American Lung Association's *State of the Air: 2006* reported that air quality in the United States improved significantly between the years 2002 and 2004, with a substantial drop in unhealthy days (particularly for ozone pollution). Fewer counties received failing grades than in previous years. The main reason was the installation of controls on NOx emissions from electric utilities in 11 eastern states in 2004.

But there is also bad news. In spite of these improvements, the Ontario Medical Association estimated that in 2005 over 5,800 deaths in Ontario were caused by bad air (Ontario College of Family Physicians, 2005). Toronto's Public Health Department reached the same conclusion, estimating that air pollution kills about 1,000 per year in Toronto. And the auditor-general of Canada agrees, stating that air pollution can be linked to 5,000 premature deaths each year in 11 major Canadian cities. This is an incredible death toll, and places an enormous burden on the health-care system, increasing emergency room visits and cardiac and respiratory hospital admissions. In Ontario, for example, there are 64,000 emergency-room visits each year because of exposure to smog (Ontario College of Family Physicians, 2005).

Data from the United States is equally chilling. In spite of improvements, the American Lung Association finds that the effects of unhealthy air nationwide are very serious. They report that nearly half of the US population (140.5 million) lives in areas with unhealthy levels of ozone; more than one in five Americans (64.3 million) live in areas with unhealthy levels of short-term particle pollution; nearly one in five people in the United States (53.1 million) live in an area with unhealthy year-round levels of particle pollution. The most alarming statistic is that nearly 15 percent of the US population (42.5 million) lives in counties where all three levels are unhealthy.

According to the American Lung Association (2006), scientists have estimated that the number of deaths in the United States associated with air pollution ranges from 50,000 to 100,000 per year. For every 75 of these deaths, there are 265 hospital admissions for asthma and 240 non-asthma respiratory admissions; 3,500 respiratory-emergency doctor visits; 180,000 asthma attacks; 930,000 restricted activity days; and 2 million acute respiratory symptom days. The cost to the health-care system is enormous. As an American Lung Association official stated, "Evidence is mounting each year underscoring just how dangerous air pollution really is. The more we learn, the more critical cleaning up the air becomes."

AIR POLLUTION AND CENSORSHIP

Air quality is not only killing people, it is also taking a toll on freedom of speech. In 2000, the US House of Representatives voted to block the Environmental Protection Agency from naming cities that are out of compliance with the Clean Air Act, and that have high levels of ground-level ozone and smog. Apparently, they feared that labeling certain communities as having polluted air would discourage economic investment—an incredible action for a country that considers itself a bastion of democracy and freedom of speech (Lazaroff, 2000).

And this picture is even gloomier in developing countries. The most polluted places in the world are the proliferating megacities of developing countries; in fact, there are more than 15 cities in the world with populations of more than 10 million. Beijing, Mexico City, and New Delhi all have particulate matter levels above 400 micrograms per cubic meter (denoted $\mu g/m^3$), which is more than eight times the levels found in Canada, the United States, or the United Kingdom, and far exceeds the World Health Organization's recommended threshold levels of 50 to 100 $\mu g/m^3$ (Lomborg, 2001).

Although China is enjoying a booming economy—which is bringing prosperity to hundreds of millions—it is also suffering from serious air pollution. This critically dangerous situation is caused largely by coal, which provides two-thirds of the country's energy, as well as the increasing use of cars. Only 1 percent of China's urban population breathes air that is considered safe. The European Union stipulates that for particulate matter less than 10 microns in diameter (known as PM_{10}) an annual average concentration above 40 micrograms is unsafe. The United States allows concentrations of 50. In 2006, according to the Chinese National Bureau of Statistics, Beijing's average PM_{10} level was 141. This airborne pollution takes a toll. In 2007, a World Bank study reported that outdoor air pollution was causing 350,000 to 400,000 deaths a year in China, while indoor pollution contributed an additional 300,000 deaths (Kahn and Yardley, 2007). Progress has a price.

The Pollutants

The World Health Organization (1992), the World Bank, the Organization for Economic Co-operation and Development, and numerous other agencies have identified the following six compounds as being key constituents of air pollution and smog:

- ozone (O_3)
- particulate matter (PM)
- sulfur dioxide (SO_2)
- nitrogen oxides (NOx)
- carbon monoxide (CO)
- lead (Pb)

Often hydrocarbons are also included in this list. Environment Canada has established National Ambient Air Quality Objectives, guidelines for provincial and federal government regulators that have no legal force. There are three levels of objectives:

- The *maximum desirable level* is the most stringent and is the long-term goal.
- The *maximum acceptable level* is intended to provide adequate health protection.
- The *maximum tolerable level* is set so that if air quality exceeds this level action is required to protect public health.

In the United States, the compounds listed above are known as the six "criteria pollutants": as such they are the only air pollutants for which the

US Environmental Protection Agency (EPA) is required to set standards, as mandated in the US Clean Air Act (amended in 1990). These standards are formally known as the National Ambient Air Quality Standards and are issued in two levels: primary and secondary. Primary standards are more stringent and are meant to protect public health, including the health of "sensitive" populations such as asthmatics, children, and the elderly. Secondary standards, which are less stringent, aim to protect public welfare, including guarding against decreased visibility and damage to animals, crops, vegetation, and buildings (EPA, 2002).

Developing regulations and setting standards requires an understanding of how and at what concentrations these pollutants can harm human health and damage the environment. In practice this is difficult because: these substances have many different chemical and physical forms, each with different consequences; their environmental pathways are complex; and often, instrumentation is not available to collect and measure the tiny quantities of pollutants that are present. As ongoing research improves our understanding of these pollutants, standards will need to be revised. The American Lung Association, for example, has recently concluded that small particulate matter is more harmful to health than was previously thought, and that its standards should be made more stringent. The US EPA is required by law to review their standards every five years.

It is worth taking a closer look at the main contaminants in smog.

Ozone

If you have ever been caught outside in a thunderstorm, you may have noticed an unpleasant, metallic smell, caused by a gas that results from lightning strikes. In 1840, the Swiss chemist Christian Schönbein named this gas "ozone" from the Greek word *ozein*, meaning "to smell".

Ozone, the major ingredient in smog, is a secondary pollutant formed by a complex interaction between NOx and hydrocarbons, which act together in the presence of sunlight. These photochemical reactions produce photochemical smog.

Nitric oxide (NO), an ingredient in the formation of ozone, is not particularly damaging in itself. But when it forms nitrogen dioxide (NO_2), it takes on a bitter smell and the whiskey-brown color that gives smog its distinctive appearance. Then an additional reaction may take place. Energy from sunlight (in the wavelength 290–420 nanometers; a nanometer is one billionth of a meter, denoted by scientists as 1×10^{-9} m) splits O (oxygen) from NO_2:

$$NO_2 + sunlight \longrightarrow NO + O$$

This reaction is primarily responsible for initiating the reactions that result in photochemical smog. The oxygen atoms, once formed, react very quickly with molecular O_2 to make the gas ozone, O_3 (Dickson and Quickert, 1975).

Both water vapor and carbon monoxide appear to contribute significantly to these photochemical reactions. Hydrocarbons react with ozone, oxygen atoms, and other compounds to form a wide variety of products of varying toxicity and corrosiveness; these include nitrogen dioxide, hydrogen peroxide, formic acid, ketones, aldehydes, and nitric and nitrous acids. The products of most interest are the peroxyacetyl nitrates (PANs) which are toxic to plants and cause eye irritation.

Ozone and most of these products are oxidants whose toxicity arises because they readily strip electrons away from other molecules (a process called "oxidation"), causing chemical reactions that can disrupt key structures and functions within cells. Because all these oxidant concentrations typically increase and decrease at the same time, usually only ozone levels are reported.

The greatest concentrations of ozone occur on hot summer days. These concentrations reach a maximum in the early afternoon when the sun is strongest and the air is filled with exhaust fumes from cars, buses, trucks, and power plants. The ozone formed in this manner resides mainly in the lower 1.5–2 km (0.9–1.2 mi.) of the atmosphere and is called ground-level ozone. It is not to be confused with the ozone layer found high in the stratosphere, which serves a very beneficial purpose in protecting life from the sun's harmful UV rays. Ground-level ozone has long been thought of as a local, urban problem; however, we now know that it can travel relatively long distances of up to 800 km (500 mi.) (United Nations, 2002).

Ground-level ozone is more poisonous than cyanide or strychnine, because it is highly reactive. The World Health Organization recently stated that there is no safe level of human exposure to ground-level ozone; in other words, any amount of ozone, no matter how small, is harmful. Studies with laboratory animals and a few experiments on humans have shown that ozone affects primarily the respiratory system, decreasing lung capacity and causing chest pain, coughing, and lung and nasal congestion. It also causes drowsiness, nausea, headache, and inability to concentrate. People with lung and heart conditions, including asthmatics and those with chronic obstructive pulmonary disease, are the most susceptible. Everyone, however, is affected to some degree.

Ozone also harms agriculture and forests. It damages foliage, with lesser effects on growth and crop yield. Tobacco and grape leaves become stippled with dark spots, and yields from orchards and vineyards are decreased. Experiments show that ozone also affects cucumbers, onions, and Christmas trees. Greenhouse experiments show that ozone decreases a plant's ability to photosynthesize, with ozone concentrations greater than 50 ppb for eight hours causing visible leaf damage. Scientists concluded that damage to non-woody vegetation can occur when ozone concentrations exceed 50 ppb for one hour or more (Young, 1977). It is estimated that this concentration is exceeded in about one out of three days in the summer in the midwestern and eastern United States. Losses to major crops such as peanuts, soybeans, wheat, and corn often exceed 10 percent, while losses to beans and tomatoes can be higher than 20 percent (Harte et al., 1991). The cost of agricultural crop damage and reductions in commercial forest yields caused by ozone in the United States is estimated at $500 million per year (United Nations, 2002). In Canada, there is also cause for concern. Ozone concentrations in southern Ontario and Quebec often peak at 160 ppb during summer.

IS IT REALLY HEALTHIER AT THE COTTAGE?

Trees growing in the countryside suffer more from air pollution than trees growing in major cities, claims American ecologist Jillian Gregg, who planted cloned eastern cottonwoods in New York City and in nearby rural areas. Her study, published in the high-profile science journal *Nature*, revealed that as a result of ozone concentrations, trees grow more slowly in the countryside. The reason is that it takes time for the urban pollution of VOCs and NOx to "cook" in hot temperatures and sunshine and create ozone. But by this time, the air mass has moved out of the city. A second factor is that cities have higher concentrations of nitric oxide, a component of smog that destroys ozone.

The findings apply to all major North American cities, states Dr. Gregg. Her research is supported by Tom Hutchinson of Trent University, Ontario. His measurements of ozone levels in the cottage country around the Kawartha Lakes show that they are double the concentrations in Toronto (McIlroy, 2003).

TABLE 4-1: Canadian National Ambient Air Quality Objectives for Ozone (Environment Canada, 1990)

Time Frame	Maximum Desirable (ppb)	Maximum Acceptable (ppb)	Maximum Tolerable (ppb)
1-hr average	50	82	150
24-hr average	15	25	–
1-yr average	–	15	–

Table 4-1 shows the Canadian National Ambient Air Quality Objectives for ozone (Environment Canada, 1990). Canada has set a target that by 2010 ozone levels will not exceed a one-hour maximum acceptable level of 65 ppb (United Nations, 2002). This may be a difficult goal to reach, as annual ambient levels of ground-level ozone increased by 33 percent between 1974 and 2001. In 2001, they registered 19.5 ppb for the year, 30 percent above the acceptable standard with less than 10 percent of individual stations recording mean readings below this level. Data on the percentage of stations with readings exceeding short-term concentration goals also indicate that ozone levels are increasing. The percentage of stations reporting at least one reading above the maximum desirable one-hour standard increased from 95.1 percent in 1977 to 97.6 percent in 2001. However, the percentage of stations exceeding the maximum acceptable standard fell from 78 percent to 47 percent (Brown and Palacios, 2005).

In the United States, the national ozone standards were lowered in 1997 (EPA, 2002).

Table 4-2 shows the US National Ambient Air Quality Standards for ozone (EPA, 2002). These are both primary and secondary standards. Although the one-hour concentrations of ozone in the United States declined by 30 percent between 1978 and 1998, the levels are still dangerously high

TABLE 4-2: US National Ambient Air Quality Standards for Ozone (EPA, 2002)

Time Frame	Standard (ppb)	Primary (P) or Secondary (S)
1-hr average	120	P and S
8-hr average	80	P and S

(United Nations, 2002), and, as noted earlier, nearly half of the US population lives in areas with unhealthy levels of ozone.

THE NEVER-ENDING ENVIRO–ECONOMY WAR

In March 2008, another battle in the endless conflict between the environment and the economy took place. The US Environmental Protection Agency (EPA) announced a new revised ozone standard of 75 ppb (over an 8-hour average), more than a decade after it was last updated. The standard sets limits on the amounts of NOx and other chemical compounds that are allowed to come out of vehicles, manufacturing facilities, and power plants across the nation. The EPA's Clean Air Scientific Advisory Committee unanimously recommended that the agency set the standard no higher than 70 parts per million (ppm)— and should in fact consider setting the level as low as 60 ppm. Many public health officials and health and environmental organizations were pushing for a more protective standard of 60 ppb, especially as the EPA's own scientific evidence indicates that people can be harmed at levels as low as 40 ppb.

But big industries mounted an intense lobbying effort to block the tougher standards. Groups representing manufacturers, automakers, electric utilities, grocers, and cement makers met with White House officials in an effort to keep the ozone standard unchanged. They argued that a more stringent standard would require costly pollution control equipment and ultimately harm the economy. For example, with a tougher standard, more US cities and counties would be out of compliance, and could face substantial economic penalties, including the loss of federal funds.

Economic interests apparently won yet another skirmish in the ongoing conflict between money and health.

It is a concern that the concentrations of ground-level ozone remain high in North America; this contaminant is one of two that are particularly damaging to health. The other is particulate matter.

Particulate Matter

The air in all major cities is filthy with small particles. When you wash your hands after a walk (even on a clear day) in New York, Chicago, Toronto, or Mexico City you may find the water that runs off is gray with dirt. After a rainfall the air feels fresher; the air is indeed cleaner, because rain is a good scrubber and removes particulates from the atmosphere, if only temporarily. The damaging role that particulates play in air pollution, however, has only recently begun to be fully appreciated.

Particulate matter (PM) is the general term used for a wide variety of microscopic solid particles and liquid droplets in the air including aerosols, smoke, fumes, dust, ash, and pollen (Environment Canada, 1998). Particulate matter is one of the most dangerous air pollutants and also the most damaging. It is estimated that it is responsible for about 90 percent of the economic cost of air pollution (Lomborg, 2001).

The size of the particle is the key factor. Larger particles do not stay airborne very long. It has also been well established that only particles with a diameter of less than ten microns (about one-tenth the thickness of a human hair) enter the thoracic region of the lung. So air pollution studies generally focus on particles of this size, which are designated as PM_{10}. Particles are further subdivided into coarse fractions (10 to 2.5 microns) and fine fractions (less than 2.5 microns, $PM_{2.5}$) because they produce different health effects. The coarse fraction commonly originates both from natural sources (such as wind-blown dust) and human activity. It tends to settle out of the air within a few hours to a few days.

Fine particulates, $PM_{2.5}$, are also known as "respirable particles". They are more hazardous than the coarser fraction of particulate matter because they penetrate more deeply into the respiratory system. Furthermore, the fine particles are better than large particles at attracting other toxic substances. $PM_{2.5}$ is largely made up of sulfate and nitrate particles, elemental and organic carbon, and organic compounds originating mainly from human-caused pollution. Sulfate and nitrate particles are created from sulfur dioxide and NOx respectively, by chemical reactions in the atmosphere. Because these fine particles can stay aloft for days or weeks, they can travel long distances. During episodes of widespread air pollution in southern Ontario, for example, it is estimated that more than 50 percent of $PM_{2.5}$ comes from the United States.

It appears that particulate matter may be the primary cause of air pollution-related mortality. It is estimated that particle pollution in the United States causes about 135,000 deaths per year (Lomborg, 2001). This is an enormous number—more than triple the number of deaths from car accidents in the United States (42,642 in 2005; car-accidents.com, 2008).

The behavior of particles in the human body is the subject of considerable research. The particles enter the lungs and cause changes in normal function, irritate the bronchi, and alter pH values. Smaller particles can enter alveoli, the air cells of the lungs. Research indicates that fine particulate molecules may be able to move directly through the lung lining and into the bloodstream (Van Dongen, 2002). Increasingly, evidence points to the smallest particles, $PM_{2.5}$, causing the most damage: they may enter the smallest crevasses in the lungs where oxygen is absorbed. Researchers have not identified a minimum concentration of $PM_{2.5}$ below which no damage is caused. As a result, North American standards have been tightened, and further tightening is expected.

The concentrations of PM_{10} in Canadian urban centers are generally in the range of 20–30 micrograms per cubic meter ($\mu g/m^3$), with episodic maximum values reaching 100 $\mu g/m^3$ and concentrations at a few sites occasionally nearing 180 $\mu g/m^3$. Concentrations of $PM_{2.5}$ in Canadian urban centers generally range from 7–20 $\mu g/m^3$ with the highest values found, as might be expected, in the Windsor–Quebec City corridor and near Vancouver, where emissions of the precursors NOx and sulfur dioxide are high.

The Ontario Ambient Air Quality Criteria for PM_{10} is 50 $\mu g/m^3$, averaged over a 24-hour period. Canada has set a goal of reducing $PM_{2.5}$ by 2010 to 30 $\mu g/m^3$, averaged over a 24-hour period. The objective for total suspended particulates is 70 $\mu g/m^3$ for an annual average (Tollefson et al., 2000). US National Ambient Air Quality Standards for particulate matter (EPA, 2002) are shown in Table 4-3.

PM concentrations in both Canada and the United States have decreased since 1955, because sulfur dioxide emissions have been reduced significantly by decreasing the use of high-sulfur coal, adding scrubbers to power plants, improving energy efficiency, and introducing catalytic converters on cars (Lomborg, 2001). But in spite of these improvements,

TABLE 4-3: US National Ambient Air Quality Standards for Particulate Matter (EPA, 2002)

Time Frame	PM$_{10}$ Standard (μg/m^3)	PM$_{2.5}$ Standard (μg/m^3)	Primary (P) or Secondary (S)
24-hr average	150	65	P and S
1-yr mean	50	15	P and S

PM levels are still unacceptably high and, with the recent aggressive American programs to expand coal-fired energy production, they may very well start to rise in the future.

Sulfur Dioxide

Sulfur dioxide (SO$_2$), along with some of the products of its reactions such as sulfates, is an important contaminant that has caused some of the worst pollution episodes in history (the 1952 London smog, for example). Sulfur dioxide, a colorless gas that smells like burnt matches, is formed under certain conditions: when fuel containing sulfur—mainly coal and oil—is burned; when ores containing sulfur are processed by smelting; and during other industrial processes such as steel-making.

Sulfur dioxide reacts with oxidants in the atmosphere and transforms from a gas into solid particles of sulfate and sulfuric acid, both of which are more hazardous than the original sulfur dioxide. Short-term exposure of asthmatics and other sensitive individuals to elevated sulfur dioxide levels during moderate exercise may result in reduced lung function, often accompanied by chest tightness, wheezing, or shortness of breath. Longer-term exposures to sulfur dioxide together with high levels of particulate matter may lead to (among other ailments) respiratory illness, changes in the lungs' defense mechanisms, and aggravation of existing cardiovascular disease. Individuals with cardiovascular disease or chronic lung disease, as well as children and the elderly, are particularly susceptible.

In the atmosphere, sulfur dioxide oxidizes to sulfur trioxide, which—when combined with water vapor—transforms readily to a sulfuric acid mist or particles. This can be transported long distances by prevailing wind currents and can cause significant environmental damage. (Acid rain is discussed in Chapter 6.) Sulfur dioxide is also a major precursor to PM$_{2.5}$, another significant health concern.

TABLE 4-4: Canadian National Ambient Air Quality Objectives for Sulfur Dioxide (Environment Canada, 1990)

Time Frame	Maximum Acceptable (ppb)	Maximum Desirable (ppb)
1-hr average	340	170
24-hr average	110	60
Annual average	20	10

Table 4-4 shows Canadian National Ambient Air Quality Objectives for sulfur dioxide (Environment Canada, 1990). Sulfur dioxide levels in Canada have improved dramatically over the past three decades. The ambient annual national mean for sulfur dioxide decreased by 72 percent between 1974 and 2001. These improvements occurred over two distinct periods: rapid improvement during the first ten years followed by a second phase of slower improvements. This is probably the result of addressing the most serious sources of pollutants first, followed by the less significant ones later.

Table 4-5 shows US National Ambient Air Quality Standards for sulfur dioxide (EPA, 2002). Between 1988 and 1997, sulfur dioxide concentrations in the United States decreased 39 percent and sulfur dioxide emissions decreased 12 percent, due largely to controls implemented under the EPA's Acid Rain Program. Between 1996 and 1997, national sulfur dioxide emissions increased 3 percent, primarily the result of increased demand for electricity.

TABLE 4-5: US National Ambient Air Quality Standards for Sulfur Dioxide (EPA, 2002)

Time Frame	Standard (ppb)	Primary (P) or Secondary (S)
3-hr avg	500	S
24-hr avg	140	P
Annual mean	30	P

Nitrogen Oxides (NOx)

"NOx" is shorthand for all the oxides of nitrogen including nitric oxide (NO), nitrogen dioxide (NO_2), nitrogen oxide (N_2O), nitrogen tetroxide (N_2O_4), dinitrogen trioxide (N_2O_3), and nitrogen pentoxide (N_2O_5). Because nitrogen (78 percent) and oxygen (21 percent) are the most abundant gases in the atmosphere, oxides of nitrogen form quite readily whenever high-temperature combustion takes place. The primary product is nitric oxide, but this quickly converts to nitrogen dioxide, which is the main pollutant of concern (although all the others are also formed, in much smaller quantities). NOx are produced by living organisms, lightning, volcanoes, and forest fires. The main technological sources of NOx are motor vehicles, power plants, and high-voltage transmission lines. Of these, the transportation sector is responsible for some 60 percent of NOx emissions in Canada and 53 percent in the United States (United Nations, 2002).

NOx cause health and environmental damage both directly and indirectly as a precursor to the creation of ozone. In addition, nitrogen dioxide mixes with fog, rain, and snow to produce acid rain and particulates. It is yellowish-brown with a distinctive odor that becomes noticeable in concentrations as low as 1–3 ppm.

Nitrogen dioxide, the most toxic of the NOx, is a deep lung irritant that damages the cells lining the lung. Unlike many other toxic compounds, nitrogen dioxide causes no symptoms until high exposure levels are reached. At levels of 5–10 ppm, NOx cause eye and nose irritation in some people. Higher concentrations can cause bronchitis and aggravate other lung disorders. Prolonged exposure to 10–40 ppm can have serious health effects. It also reduces resistance to bacterial and viral infections, suppresses growth of vegetation, and can corrode metals (Environment Canada, 1990). NOx are less dangerous than other pollutants such as ozone and particulate matter.

The amount of NOx in the atmosphere has decreased in recent years, primarily due to the introduction of catalytic converters in cars. From 1974 to 2003, ambient levels of nitrogen dioxide in Canada decreased by over 34 percent. Annualized concentrations of nitrogen dioxide have steadily declined in Canada, remaining below the maximum desirable objective of 32 ppb (Brown and Palacios, 2005).

The Canadian National Ambient Air Quality Objectives for nitrogen dioxide (Environment Canada, 1990; Tollefson et al., 2000) are shown in

TABLE 4-6: Canadian National Ambient Air Quality Objectives for Nitrogen Dioxide (Environment Canada, 1990; Tollefson et al., 2000)

Time Frame	Maximum Desirable (ppb)	Maximum Acceptable (ppb)	Maximum Tolerable (ppb)
1-hr average	–	210	530
24-hr average	–	110	160
Annual average	32	53	–

Table 4-6. The US National Ambient Air Quality Standard for nitrogen dioxide, based on an annual arithmetic mean, is 53 ppb. This is both a primary and secondary standard (EPA, 2002).

The United States has set a national cap for NOx for the five summer months (May to September). As already in place for sulfur dioxide (see Chapter 6), utilities will be assigned NOx emission allowances with the ability to trade or save unused credits.

Carbon Monoxide

Carbon monoxide (CO) is a colorless, tasteless, odorless, and poisonous gas produced by incomplete combustion. The main sources are vehicles (more than 66 percent), cigarettes, and the improper use of indoor cooking and heating appliances. Lethal concentrations can occur, particularly in enclosed spaces such as garages.

When inhaled, carbon monoxide deprives the body of oxygen by taking oxygen's place in binding with the blood's hemoglobin. It also hinders the release of oxygen from the blood to the body's tissues. Exposure to carbon monoxide can occur in rush-hour traffic, busy parking garages, or enclosed spaces. Symptoms include headache, dizziness, drowsiness, and nausea. The body organs affected are those that rely the most on a steady supply of oxygen: the heart and the brain. A fetus may also suffer oxygen deprivation when its mother is exposed to carbon monoxide.

Tables 4-7 and 4-8 show the national standards for carbon monoxide for Canada and the United States, respectively. In Canada, ambient levels of carbon monoxide have been dramatically reduced over the past three decades. From 1974 to 2001, ambient levels of carbon monoxide dropped by 83 percent, and are well below the national objectives. Over the same period, total vehicle registrations increased more than 30 percent. As with NOx, the drop in ambient carbon monoxide is mainly the result of catalytic converters, which remove over 85 percent of carbon monoxide from vehicle emissions (Lomborg, 2001). This decrease in ambient carbon monoxide represents Canada's second-most effective campaign at eliminating pollutants, second only to the decrease in lead (Brown and Palacios, 2005).

TABLE 4-7: Canadian National Ambient Air Quality Objectives for Carbon Monoxide (Environment Canada, 1990; Tollefson et al., 2000)

Time Frame	Maximum Desirable (ppm)	Maximum Acceptable (ppm)	Maximum Tolerable (ppm)
1-hr average	13	31	–
8-hr average	5	13	17

TABLE 4-8: US National Ambient Air Quality Standards for Carbon Monoxide (EPA, 2002)

Time Frame	Standard (ppm)	Primary (P) or Secondary (S)
1-hr average	35	P
8-hr average	9	P

Lead

Lead is a soft, silvery-white or gray, very malleable, dense metal. It is a poor conductor of electricity. These properties, as well as its high durability and resistance to corrosion, have meant that lead has been widely used since it was first mined 5,000 years ago. The Romans made pipes in their water systems from lead. Today, lead is used in a wide range of applications including roofing, ammunition, coverings for electric cables placed in the ground, linings for water pipes, shielding for electromagnetic radiation and radioactivity, sound baffles in buildings, and structures for corrosive substances. The biggest use of lead is in manufacturing storage batteries. For many years, lead was an important additive to gasoline and to paint, although these practices have now ceased. The mineral galena—lead sulfide (PbS)—is the major source of lead production throughout the world.

As a result of its widespread use, lead is commonly found in air, water, soil, and food. It enters the air mainly from vehicle exhaust, burning coal in power plants, and incinerating solid wastes. It is extremely toxic and can cause severe health effects even at relatively low levels in the body, including irreversible brain damage and injury to the blood-forming systems. Lead affects the nervous system, the production of blood cells, the kidneys, the reproductive system, and behavior (Harte et al., 1991). Symptoms include pallor, cramps, vomiting, listlessness, coma, and death. Unborn children are particularly susceptible, with studies showing an increase in miscarriages if either spouse has been exposed to lead in the workplace. Lead exposure can reduce men's fertility and doubles the risk of women giving birth to children with learning disabilities. The World Health Organization reported that in European children aged from one to four, approximately 150,000 years of healthy living were lost due to lead in air pollution (WHO, 2004). In the United States, it is estimated that 890,000 children between the ages of one and five have levels of lead in their blood high enough to affect intelligence, growth, and development (Safe Kids, 2004). Children between one and two are at the greatest risk from lead poisoning.

Some lead intake can come from the ingestion of lead-based paint. Although it has been banned since the 1950s, many older buildings still have some lead paint on trim, walls, and doors. About 90 percent of lead intake, however, comes from vehicle gasoline. A phase-out in the United States began in 1973 and was completed in 1986; now all gasoline in that country is lead-free. Other countries have had similar programs. Monitoring stations throughout the United States show that lead in air has decreased by about 97 percent since 1973—an enormous improvement (Lomborg, 2001).

Canada has not established objectives for lead concentrations in air. The US National Ambient Air Quality Standards for lead, based on a quarterly average, is 1.5 $\mu g/m^3$, which is both a primary and secondary standard (EPA, 2002). Thanks to the phasing out of lead in gasoline, the average levels of lead measured in urban air have decreased significantly and are far below this standard, which was established in 1978. In 2008, the US EPA initiated steps to lower the standard to 0.1–0.5 $\mu g/m^3$ (EPA, 2008a).

Hydrocarbons

Ambient air levels for most hydrocarbons are not measured routinely because the large number and variety of them—there are several thousand compounds—present in the air at any one time make it virtually impossible. Furthermore, hydrocarbons are not criteria air pollutants in either the United States or Canada.

Some hydrocarbons, however, are regularly monitored given their significance to human health. These include: VOCs, a key ingredient in the creation of ozone; benzene, which causes cancer; and methane, which contributes to global warming. About 50 percent of VOCs in the air are caused by burning gasoline. These are added to gasoline to help it burn evenly. Other sources of VOCs include commercial activities such as printing and dry cleaning.

Polycyclic aromatic hydrocarbons (PAHs), a group of over 100 related chemicals, are produced when organic materials such as coal, oil, gas, wood, garbage, and tobacco are burned, especially with insufficient oxygen. Benzo(a)pyrene is the most toxic of the PAHs. Forest fires are the largest natural source, and wood combustion in residential fireplaces is also a major contributor (Environment Canada, 1990).

FIGURE 4-1

Canadian National Air Pollution Surveillance station in Alberta (Environment Canada)

Monitoring

We have seen how understanding the distribution and effects of air pollution depends on accurately monitoring the amounts of pollutants in the air, in many locations around the world. Ground-based measurement with direct air sampling is the mainstay of air-quality studies. In Canada, the National Air Pollution Surveillance (NAPS) network was established in 1969 with the focus on urban centers. Provinces operate and maintain the monitoring stations, while the federal government supplies the bulk of the instruments and coordinates the operation of the network to ensure uniform standards. The network has grown steadily and by 2002 included 800 instruments at 253 stations in 156 urban centers, with an ongoing commitment to expand the network. Figure 4-1 shows a typical NAPS monitoring station. The NAPS network provides data that are used to:

- conduct research into the causes of air contaminants, how they are transported through the atmosphere, and their health and environmental effects
- support the setting of regulatory standards and to monitor the results of regulations
- predict and provide air-quality information to the public (Environment Canada, 1990; Turle, 2003)

The number of pollutants being monitored by NAPS continues to grow. Initially, only carbon monoxide, ozone, total suspended particles, sulfur dioxide, NOx, and lead were monitored. Since then, the following have been added: PM_{10}; $PM_{2.5}$; over 50 elements; sulfate (SO_4); nitrate (NO_3); VOCs such as benzene, chlorinated hydrocarbons, and carbonyls; semivolatile organic compounds such as polychlorinated aromatic hydrocarbons, dioxins, and furans. In total, over 270 chemical species and many individual chemicals are measured. It is a large and complex undertaking.

In the 1970s, shortly after the NAPS network was established, it was expected that regulatory controls would cause major declines in airborne pollutants. The network showed that up to the mid-1980s, the concentrations of sulfur dioxide, carbon monoxide, total suspended particulates, and lead were indeed decreasing. Since then, however, the decreasing trends have slowed and levels of some pollutants, such as NOx and ozone, have increased. At the same time, health studies have shown that some pollutants, such as $PM_{2.5}$, have a much greater impact on human health than previously thought.

These results, combined with the growing interest of the public in environmental matters, show the value of the NAPS network.

A large and complex network has been established in the United States to monitor air pollution, consisting of national, state, local, and industrial air monitoring stations. Ozone, for example, is measured at 616 state and local air monitoring stations, 194 national stations, and over 250 other sites. The massive amounts of data produced by this network are collected and coordinated by the EPA through their Air Quality System database and can be viewed on the EPA website.

Agreements between the United States and Canada have required monitoring, given that these pollutants can travel long distances.

Instruments for Monitoring Pollutants

To perform the complex measurements of gases and minute airborne particles, dozens of different high-technology instruments are used both in the field and in the laboratory. Significant funding is going into developing instruments that can measure the tiny concentrations—parts per billion or smaller—of an ever-increasing array of pollutants. In addition, new instruments and measurement techniques must continually be developed to detect the thousands of new synthetic chemicals introduced into the environment each year.

A PEEK INTO THE HIGH-TECH WORLD OF CHEMICAL ANALYSIS

Analytical chemists have devised increasingly sophisticated tools to identify and measure atmospheric pollutants. At Environment Canada's Environmental Technology Centre, Ottawa, I learned, for example, that dioxins and furans are a family of 210 closely related organic compounds whose full names are "polychlorinated dibenzodioxins" and "polychlorinated dibenzofurans".

Air samples are collected using a high-volume filter and polyurethane foam cartridge. The dioxins or furans are extracted from the cartridge and dissolved in an organic compound. This is filtered through sodium sulfate and then purified by running the solution through a series of acid-base columns followed by a basic alumina column. The resulting concentrate is injected into a "gas chromatograph" column designed for base neutral compounds. A gas chromatograph is essentially a very long, thin tube. As the sample travels, it interacts with the tube and the sample components are gradually separated. By choosing appropriate tube materials, molecules that differ only slightly in their chemical or physical properties can be separated.

As the separated sample components emerge from the chromatograph, they are injected into a high-resolution "mass spectrometer", which produces charged particles—ions of the original molecules. By accelerating the ions in a magnetic field, the components are separated and identified according to their mass to electric-charge ratios.

Extremely low levels of about 1 femtogram (one part in 10^{18}, that is, one part in a billionth of a billionth, see Scientific Notation on page vii) of dioxin or furan per cubic meter of air can be detected. Analyzing air for dioxins and furans is extremely expensive and there are only a handful of laboratories with this capability in Canada.

Just maintaining the equipment in the field is a complex undertaking given the large number of instruments, their high-tech fragility, and the fact that they must operate continuously in all kinds of weather. Furthermore, all the instruments must be properly calibrated so their results can be compared with confidence across the country and to ensure that trends over time are meaningful.

And finally, the enormous amounts of information produced by the network must be received, decoded, and converted to appropriate units; organized and stored in a retrievable manner; analyzed by scientists; and reported to the public and other interested parties. The data management alone is a gargantuan task.

In recent years satellites have been playing an increasingly important role in helping scientists understand pollution on a regional and global scale. Using the US space agency's Aura satellite, for example, launched in 2004, they can look down at the troposphere and use the Ozone Monitoring Instrument (OMI) and other equipment to build a daily picture of air quality. OMI measures nitrogen dioxide, the precursor to smog, which comes from motor vehicle exhausts, power plants, and industry.

By following the development and spread of nitrogen dioxide, the OMI can be used to help forecast where problem air might develop. Long-term tracking of the gas can also identify emission hotspots. OMI can also detect formaldehyde, sulfur dioxide, small particles, and tropospheric ozone.

The Air Quality Index

In response to increasing public interest in air quality, governments have sought ways to communicate environmental information to the public. As growing cities produce more and more smog, there is a particular need for an index that describes air quality and allows officials to warn of impending poor air quality. Such an index should be not only easy to understand, but also accurate and comparable between communities. And, it should be expandable to include additional pollutants when necessary.

The Air Quality Index (AQI) was first developed in Canada in 1976, and is now used by most provinces and the federal government. The AQI is still evolving and its definition and use vary somewhat from jurisdiction to jurisdiction (Hewings, 2001).

In Ontario, the AQI was originally based on measurements of the five most common pollutants: sulfur dioxide, ozone, nitrogen dioxide, total reduced sulfur compounds (these cause bad odors), and carbon monoxide. Every hour, the concentration of each pollutant is converted into a whole-number value. The pollutant with the highest value becomes the AQI reading. The higher the value, the worse is the air quality, as seen in Table 4-9.

Because particulate matter is highly toxic, in 2002 Ontario upgraded its Air Quality Index to include $PM_{2.5}$. This yields a more realistic view of

TABLE 4-9: Air Quality Index

AQI Value	Category	Description
less than 32	good	The air quality is considered relatively good.
32–49	moderate	There may be some adverse effects on very sensitive people.
50–99	poor	There may be some short-term adverse effects on human or animal populations, or some significant damage to vegetation and property.
100 or over	very poor	There may be adverse effects on a large proportion of those exposed.

air quality and was expected to increase the number of air quality advisories issued in a year by about 10 percent.

The calculation of the AQI number for each pollutant is based on the three federal standards (maximum desirable, acceptable, and tolerable objectives) which are plotted on a graph. The AQI value for a specific concentration of one pollutant is then read from the graph. The procedure is repeated for all six pollutants, and the highest number becomes the AQI.

An obvious shortcoming is that a single pollutant determines the AQI, with the remaining ones playing no role. Suppose that on two different days the AQIs are:

	Day 1	Day 2
$PM_{2.5}$	72	72
Ozone	70	12
CO	70	11
Reduced S	70	7
SO_2	70	9
NO_2	70	15

Clearly, Day 1 has worse air quality than Day 2, yet the same AQI of 72 would be reported both days. An inconsistency also arises if one day an AQI of, say, 68 is caused by reduced sulfur and another day the same AQI of 68 is caused by $PM_{2.5}$. The health effects of these two pollutants are different, so reporting equal AQIs for both days is not accurate.

In spite of these shortcomings, the AQI has become a valuable tool. The Canadian government is working to improve the accuracy of the AQI, particularly in terms of health risks, as well as to ensure that its use is uniform and standard across the country.

In 2006, Environment Canada and Health Canada introduced the Air Quality Health Index (AQHI), based on the health risks of a mixture of ozone, particulate matter ($PM_{2.5}$ and PM_{10}), and nitrogen dioxide (Environment Canada, 2008). The values range from 1 (very low health risk) to 10+ (very high health risk). In 2008, the AQHI was available as a pilot project in only a few locations across the country, but once it becomes available everywhere it will supplant the older AQI.

Ontario also operates a Smog Alert that provides advance warning that poor air quality caused by ozone is coming. There are two warnings:

- *Smog Watch*: there is a 50 percent chance that the AQI will exceed 50 in the next three days.
- *Smog Alert*: there is a strong likelihood the AQI will exceed 50 soon.

Smog alerts are announced with weather forecasts, are available via a toll-free telephone line, and individuals and organizations can subscribe to receive warnings by email.

When an AQI value or smog alert is issued in the media, the public may not appreciate the enormous effort required to produce it. First, the concentrations of the pollutants are measured by the NAPS network. Then the measurements are telecommunicated to a central site where they are processed each hour and AQI values are calculated. (The objectives for each pollutant, and how these are translated into AQI values, are reviewed periodically and updated as necessary by technical experts and committees.) The AQI values are combined with meteorological data to make predictions about smog. This information is then transmitted to the media and placed on websites and hotlines for the public.

In the mid-1970s the United States EPA developed the Pollutant Standards Index, which reported on air quality in metropolitan areas. In 1999, it was renamed the Air Quality Index. It is similar to the Canadian AQI in that it is based on national standards, but it ranges from 0 to 500 and incorporates the category "unhealthy for sensitive groups". It also uses color symbols so it can be interpreted visually. An important feature is that the EPA must report the AQI in urban areas with a population of more than 350,000.

Trends

Over the past few decades, the smog picture in North America has changed significantly as regulatory controls have reduced emissions of some pollutants, while other pollutants have emerged as areas of concern. Good progress has been made with sulfur dioxide, carbon monoxide, lead, and PM_{10}, which are now being regularly monitored at concentrations below the national standards established by both the United States and Canada. In particular, levels of lead and carbon monoxide in the air have decreased substantially.

Ozone is now the most pervasive air pollution problem in North America (United Nations, 2002). The main reason is that population and cities are growing, and they are crammed with cars, buses, trucks, and factories all emitting increasing amounts of smog-generating pollutants. In addition, solutions are difficult because often the underlying science is not well understood. For example, it not known which pollutant is more toxic—VOCs or NOx—and which should receive stricter regulations. The good news is that VOC emissions fell 43 percent between 1970 and 2000 in the United States; unfortunately, at the same time, NOx emissions increased by 20 percent (United Nations, 2002).

In addition, $PM_{2.5}$ has been found to be a serious problem. Both $PM_{2.5}$ and ozone continue to pose a serious health risk and their concentrations often exceed national standards.

I note, with some sadness, that according to the Air Quality Index, southern Ontario—including Toronto and the Niagara area—is the most polluted place in Canada, and possibly all of North America. Wind and weather systems channel contaminants from a large fraction of the continent into this area (Environment Canada, 1998). Southern Ontario is subject to the following major sources of pollution: cars and industry; large steel works in Hamilton, Ontario; one of the largest coal-fired power plants in North America at Nanticoke, Ontario; and the industrial midwest of the United States with much industry and coal-fired power plants. NOx and VOC emissions from the midwestern US states create ozone in the Ohio Valley which then flows into Canada; here it can contribute 30–90 percent of the total ozone (United Nations, 2002).

Solutions

Three different approaches may hold solutions to the smog problem: imposing regulatory controls, establishing marketplace credits, and encouraging people to change their lifestyles.

In Canada, a key part of the battle against smog is regulating the transportation industry by making vehicles more efficient and reducing pollutants, especially sulfur, in fuels. These regulations are discussed in the next chapter. All provinces and the federal government have agreed to an accelerated target of not exceeding ozone concentrations of 65 ppb by 2010. Ontario, for example, has an Anti-Smog Action Plan, which includes reducing emissions of NOx and VOCs by 45 percent from 1990 levels by 2015 (Ministry of the Environment, 2002).

In the United States, the Clean Air Act requires the EPA to regulate sources emitting major amounts of 188 toxic air pollutants. As of June 1999, the EPA had issued 43 regulations that would reduce emissions by more than 1 million tonnes (1.1 million tons) per year. In addition, regulations have targeted the transportation sector, one of the main sources of smog. As an initial step, the EPA identified 33 of the 188 air pollutants that pose the greatest health risks to urban areas. (Diesel fuel is being addressed separately.) In addition, regulations are being developed to reduce the pollutants of greatest concern from larger industrial sources such as chemical plants and steel mills, as well as smaller commercial and industrial sources such as dry cleaners. By 2004, regulations were in place for 29 sources, including landfills, oil and natural gas production, hospital sterilizers, paint-stripping operations, and factories that make chlorine products using mercury-cell technology. Continued research and collaboration with industry are key parts of the EPA's strategy.

In 1994 the EPA introduced an innovative method for encouraging reductions in sulfur dioxide emissions—one of the major contributors to acid rain and smog. It is a novel idea, relying on the marketplace rather than regulations. This market-based emissions-trading encourages companies to reduce their emissions far below their caps, rather than just meeting the regulatory requirements. The results have been excellent with a significant reduction in sulfur dioxide emissions. (Chapter 6 looks at this in more detail.)

LIFTING SMOG, AN OLYMPIAN FEAT

The city of Atlanta is renowned not only for its southern hospitality but also for its dynamic growth, which has resulted in congested highways, extensive urban sprawl, and a curtain of pollution. An unexpected benefit of hosting the Olympics in 1996 was a temporary lifting of the veil of smog. The level of the key smog pollutant, ozone, dropped by almost 30 percent. This decrease yielded an immediate health and economic benefit: asthma attacks in children dropped by a dramatic 42 percent, and the number of children requiring hospitalization fell by 19 percent.

Unfortunately, the asthma rates quickly rose back to their normal rates once the Olympians left the city.

Research, as reported in the *Journal of the American Medical Society*, found that Atlanta had provided massive public transportation for the Games while at the same time closing the city center to cars. The resulting reduction in vehicle emissions led to a 17-day outbreak of improved health. It also provided a clear vision of what all major cities should be aiming for—an Olympian goal indeed.

Finally, smog has become such a pervasive and insidious problem that to overcome it we must consider lifestyle changes. As the text box on the Atlanta Olympics indicates, one solution, and probably the most important, is to reduce our use of automobiles. This is not an easy task, however, since our modern way of life depends on them. If we really want to reduce smog we will need to embrace, among other measures, "smart growth", the increased use of mass transit, and hybrid cars.

Smog is a fascinating and complex problem. On the one hand it seems simple to resolve: you would only have to invoke regulations that decrease the emission of VOCs and NOx, the precursors of ozone, from tailpipes. But unfortunately, it is not so simple. Other sources also contribute to smog, such as power plants and factories that burn coal or oil and, in the process, emit sulfur dioxide, NOx, and particulate matter. The first two contribute to acid rain, and all three pollutants reduce visibility and create regional haze. As we shall see throughout this book, smog, acid rain, ozone depletion, and global warming are interlinked, sometimes directly, other times in more subtle ways. Although I describe these pollution issues in separate chapters, it cannot be overemphasized that they are interconnected.

Transportation: A Large Environmental Tire Print

Transportation is the largest source of air pollution in Canada.
—Environment Canada

The other day a friend visited and we all streamed out to admire his proud new possession. We encircled a sporty, fire-engine red sedan that gleamed in the driveway and ran our hands lovingly along the smooth slope of the hood, still warm from the throb of its 4.0 liter V-6 engine. We admired the sleek control panel, the sunroof, and the hand-stitched leather upholstery. Beside this beauty my basic little car looked as though it was ashamed and wanted to hide. Although I knew my friend could not afford this luxury, I still felt a pang of jealousy; for after all who does not dream of owning a shiny new Porsche, a sleek black Mercedes, or a powerful Corvette?

The Car—A Pillar of Society

In North America the car is more than a status symbol. It is a cornerstone of our life, allowing us the freedom to shop, work, visit relatives and friends, and travel to faraway places. Cars have enormous influence over us, largely defining where and how we live. As Henry Ford said in 1930, "The people of the United States do not own automobiles because they are prosperous. They are prosperous because they own automobiles and use them as tools to increase the range of their abilities" (Nadis and Mackenzie, 1993). It is difficult to imagine our lives without the motor vehicle.

THE CAR IS A MILESTONE IN LIFE

When older married couples were asked to list the most important milestones in the lives of their grown children, many offered the following: when their children began to walk, when they started to drive, and when they got married.

In 1956 President Eisenhower took a giant step toward making North America a car-dominated society by passing the Interstate Highway Act, which resulted in the construction of 71,000 km (44,100 mi.) of toll-free expressways crisscrossing the country (Nadis and Mackenzie, 1993). Since then, an incredible infrastructure costing untold billions of dollars has been built to support the car: millions of kilometers of roads, overpasses, cloverleafs, service stations, parking lots, traffic lights, garages, and more. Today, about 25 percent of urban space is covered by pavement. Governments continue to promote the dependence on the car by keeping gasoline prices artificially low, subsidizing road construction, and only half-heartedly supporting urban transit and railways. One of the most important factors in making society dependent on automobiles has been urban design, with its low-density housing in suburbs far removed from city centers, shopping areas, and workplaces.

The Car as Serial Killer

Yet the car has a split personality, a Jekyll-and-Hyde character, for it is also a serial killer. The US National Highway Traffic Safety Administration in 2007 reported that 42,642 deaths and 2,575,000 injuries were caused by traffic accidents in 2006. Furthermore, the economic cost of vehicle crashes in the United States in 2000 was estimated at $230.6 billion. The roads are no safer in Canada, where 2,889 people died and 199,337 were injured in traffic accidents in 2006 (CSC, 2007).

And if the car doesn't get you with the sudden squeal of tires and crunch of metal, it does so slowly with silent emissions from the tailpipe. Even on days when smog is not hanging over the city, there are still fumes that invisibly degrade the air. It may only be years later that you find yourself coughing and wheezing and perhaps dying much sooner than you should have. Few realize that more deaths worldwide are caused by air pollution (approximately three million per year) than by traffic accidents (approximately one million) (WHO, 2003).

The transportation sector leaves a very large environmental "tire print". In addition to causing significant air pollution, automobiles consume an enormous quantity of gasoline and other resources such as steel, aluminum, rubber, and glass. The story of transportation emissions and the efforts to reduce them sums up the difficulties faced by society in dealing with virtually all forms of pollution.

In its early days, in the first half of the twentieth century, the car was much dirtier than it is today, emitting a large quantity of contaminants. The deadliest of these was lead. Lead was widely used in gasoline to reduce knocking and to lubricate exhaust valves.

Thomas Midgley Jr. elegantly demonstrated, not once but twice, how little we really know about the complex interaction between the biosphere and large quantities of synthetic chemicals. Midgley combined the traits of a brilliant scientist with the showmanship of P.T. Barnum to become one of the truly memorable characters of science and engineering. After graduating from Cornell University in 1911 with a degree in mechanical engineering, he worked for the General Motors Research Corporation in Dayton, Ohio—then the mecca of technology in America. Dayton was home to the burgeoning bicycle, aviation, and automobile industries, as well as Orville and Wilbur Wright, the inventors of the first successful airplane (Cagin and Dray, 1993).

In 1916, the automobile was in its infancy and just starting to gain a toehold as a consumer product. A formidable problem at that time was that the engines knocked terribly, caused by premature explosion of gasoline. Midgley discovered that the problem was caused by the fuel and not by the firing system. He went on to experiment with over 3,000 compounds before discovering in 1921 that the additive tetraethyl lead not only removed the knocking but increased fuel efficiency and power by 25 percent. In 1923, Midgley won the American Chemical Society's Nichols Award for this work.

In 1924, the Ethyl Gasoline Corporation began to manufacture enormous quantities of tetraethyl lead. Since lead is a neurotoxin, causing symptoms including blindness, insomnia, kidney failure, palsies, and convulsions, factory workers soon began to suffer. Ethyl Corporation promptly denied that there was a health risk. It is interesting to note that similar campaigns have since denied that sulfur dioxide and nitrogen oxides (NOx) are connected with acid rain, that chlorofluorocarbons (CFCs)—another Midgley invention!—cause stratospheric ozone depletion and, more recently, that greenhouse gas emissions are in any way related to global warming (Bryson, 2003).

When the public became concerned about the health hazard of his lead-based gasoline, Midgley held a news conference in 1924. In front of the assembled press and government health officials, he vigorously washed his hands and face with the feared gasoline (Cagin and Dray, 1993). Although he himself had already developed symptoms of lead poisoning, Midgley felt that the lead additive was safe for the public because of the dilute form in which it was used.

Perhaps he was correct at that time, but he had no idea how much the population of the United States was to grow and how ubiquitous the automobile was to become. With millions and millions of tailpipes each emitting a small amount of lead, the environmental total became a serious problem. In the 1970s, lead began to be phased out.

MIDGLEY THE INVENTOR

Midgley was constantly creating inventions, including a method to extract bromine from the sea, a way of salting popcorn before it was popped, a system for controlling aerial torpedoes, and a procedure for approximating the vulcanization of rubber. He would often dumbfound his companions by stopping in the middle of a sentence, staring into the distance for a long time, and then announcing he had just made a discovery.

In 1941, Midgley received the Priestley Prize, considered the American equivalent of the Nobel Prize in chemistry. He had recently contracted polio and was confined to a wheelchair. Even with this handicap he was an eager showman and when wheeled forward to receive the prize he also brought out an engine with two glass fuel tanks on top. He waited for complete silence and then threw a switch to start the engine running with one fuel. The room reverberated with the hiccuping knock of an old jalopy. Then he threw a switch changing the engine to the other fuel. The rackety roar quickly quieted to a smooth hum. He had convincingly demonstrated the effectiveness of his discovery of the gasoline additive tetraethyl lead.

Midgley also invented CFCs, which we will see again in Chapter 8. With two major discoveries and many minor ones to his credit, Midgley was a popular and sought-after speaker and consultant. His polio barely slowed him and he was constantly on the go. His career culminated when he was elected president of the American Chemical Society in 1944, a proud moment indeed for a man with no formal training in chemistry. Midgley passed away in 1944, revered and honored, but with no inkling that his inventions of tetraethyl lead and CFCs would cause such severe damage and clearly demonstrate—not once, but twice—the dangers of introducing chemicals into the environment.

In addition to lead, automobiles emit many other pollutants. Carbon monoxide is created when the fuel mixture is too rich, with insufficient oxygen to burn all the fuel. (This is why carbon monoxide emissions are greatest when an engine is first started.) Hydrocarbons, consisting of unburned gasoline and oil vapors, are caused by incomplete combustion in the cylinders and are a major contributor to the formation of ozone and ground-level smog.

Particulates also form from incomplete combustion. Diesel engines, in particular, are notorious for high levels of particulate emission. We have all seen the black smoke coming from a truck or bus exhaust pipe as it labors up a slope or changes gears. Scientists estimate that as much as 91 tonnes (100 tons) of particulate matter are emitted in most large American cities each day. Most of the particulate matter is caused by trucks and buses (Carley and Freudenberger, 1995).

When the temperature in the engine's cylinders exceeds about $1,390°C$ $(2,530°F)$, the nitrogen (N_2) in the air combines with oxygen to form various oxides of nitrogen leading to nitrogen dioxide, the brownish gas we met earlier, which contributes to smog. Sulfur dioxide is emitted by cars because all crude oil contains some sulfur.

There are three sources of air emissions from cars. First and foremost are those from the tailpipe, which are caused by incomplete combustion and the presence of unwanted substances in the gasoline. These emissions are the most difficult to control because of the many variables involved, including composition of the fuel, air-to-fuel ratio, ignition timing, combustion temperature, cylinder shape, and so on.

The second source of emissions is the loss of gasoline vapors, known as "evaporative emissions", from the gasoline tank and carburetor. The lighter constituents of gasoline evaporate easily, especially in warm weather. These include aldehydes, aromatics, olefins, and paraffins, which can all react with air and sunlight to form ozone and smog. In fact, aldehydes are often called "instant smog" because they can form smog without undergoing photochemical changes. Evaporative emissions can account

for about 20 percent of a vehicle's total emissions. It is sobering to realize a parked car with the engine turned off can still pollute the air.

The third source of emissions is the gases from the engine's crankcase.
These are called "blowby emissions". They occur because some vapors
escape past the cylinder rings into the crankcase. The rings wear, especially as a car ages, and the cylinder walls become scored. To relieve pressure buildup, these gases were originally vented directly into the air.

Automotive Pollution Control

Two factors were to provide a huge impetus to the car industry to improve
its environmental performance. First, the late 1960s to early 1970s was a
period of social turmoil and the beginning of widespread concern about
the environment. This period saw the introduction of Earth Day; the creation of Environment Canada and the US Environmental Protection
Agency; and, particularly important to the car industry, the passing of the
US Clean Air Act in 1970. The act called for drastic—"draconian", according to the car manufacturers—reductions in automotive emissions. For
the next few years, Detroit, which until then had been unhindered in
building ever-larger, more gas-guzzling cars, was forced to rethink and
redesign ignition, carburetion, and engine systems to meet the new emissions regulations.

Then came the OPEC-induced energy crisis of 1973. This rang the
death knell for the big V-8 engines—at least temporarily—and further
spurred efforts to improve engine efficiency and reduce emissions.

Vast improvements in pollution control have been made since the early
days of the automobile. The first automotive pollution control device was
introduced in 1963 to prevent blowby emissions. Called "positive crankcase
ventilation", it circulated the gases back into the intake manifold and then
into the cylinders to be burned again. This relatively simple step virtually
eliminated crankcase emissions as a source of air pollution. In addition,
it improved engine life, without reducing power.

A major breakthrough came in 1973, when General Motors announced
it had developed a new catalytic converter that could meet the extremely
strict emission targets set by the EPA: a 90 percent reduction in hydrocarbon and carbon monoxide emissions for 1975 vehicles. This invention has proven to be one of the most effective tools in reducing car
emissions.

Since then, many other advances have been introduced including electronic fuel injection, computerized engine control, and onboard emissions diagnostics. These steps have not only reduced pollution but have also improved fuel efficiency and car performance.

THE CATALYTIC CONVERTER

Catalytic converters are canisters placed in the exhaust pipe before the muffler. They contain "catalysts", substances that promote chemical reactions without actually taking part themselves. Two rare metals, platinum and palladium, are catalysts that promote the reaction of hydrocarbons and carbon monoxide with oxygen. That is, these catalysts cause hydrocarbons and carbon monoxide to burn. The final products are water and carbon dioxide.

Catalytic converters are ceramic cores coated with a microscopically thin layer of catalyst. There are two types of catalytic converters: "monolithic honeycomb" and "pellet bed". The spaces of the honeycomb design provide an immense surface area to maximize contact between emission gases and the catalysts. The pellet bed design has many small pellets coated with the catalysts, also providing a large surface area. When the pellets become contaminated they can be easily replaced.

In 1981, a rhodium catalyst to break down NOx into ordinary nitrogen and oxygen was added to form a three-way catalytic converter. Because this reaction has to be carefully controlled, it is done in a separate chamber in the converter before the exhaust gases reach the chamber containing the platinum and palladium catalysts.

Catalytic converters do not decrease a car's efficiency or power. Today's regulatory standards require at least 160,000 km (99,400 mi.) of service before converters lose their effectiveness, a requirement that is enforced in most states and provinces by regular inspections.

In addition to its high toxicity, lead also fouls the catalytic converter, rendering it useless. So as converters were brought in during the late 1970s, lead was phased out.

Maintaining a well-tuned engine is important; in combination with a good catalytic converter, this ensures that virtually no carbon monoxide is emitted from the tailpipe. Maintaining a proper air-to-fuel ratio also minimizes hydrocarbon emissions.

NOx emissions are reduced by a method called "exhaust gas recirculation", in which a small amount of exhaust gas is recirculated back into the intake manifold to dilute the air-fuel mixture. This cools the combustion, keeping temperatures below the level at which NOx forms. This process, combined with the three-way catalytic converter, has substantially decreased the amount of NOx emitted by cars.

Evaporative emissions are controlled by sealing the fuel tank and carburetor. A pressurized gasoline fill cap and valve allow air to enter as gasoline is used up, and permit gasoline expansion and contraction with temperature change while preventing emissions from escaping. The carburetor is also sealed. Vapors are directed to a canister filled with activated charcoal, which absorbs the vapors and later releases them into the intake manifold when the engine is running.

Sulfur is a nasty contaminant in gasoline, and is not removed by catalytic converters. It not only causes sulfur dioxide emissions but also creates sulfates, which form fine particulates ($PM_{2.5}$). In addition, sulfur reduces the efficiency of catalytic converters, enabling the emission of more of the other pollutants.

Before 2002, Canada had the dirtiest gasoline in the industrial world, with an average sulfur content of 343 parts per million (ppm). During the summer smog season of 1998 Imperial Oil's refinery in Sarnia produced regular gasoline that contained 810 ppm of sulfur (Tollefson et al., 2000). In 2002, Canada announced an interim standard of 150 ppm. This benchmark was lowered to 30 ppm in 2005 (IJC, 2002).

The US EPA phased in a 30 ppm standard for sulfur in gasoline between 2004 and 2006; previously, gasoline in the United States contained about 300 ppm of sulfur.

Progress in reducing emissions is by no means ended, although further improvements are increasingly difficult and expensive.

Periodic vehicle emissions testing is required in most Canadian provinces and US states, and is becoming relatively sophisticated. Portable testing instruments can be used on the roadside to catch "gross polluters". To make the cost of meeting emission control and testing requirements easier for consumers, car manufacturers are required to provide extended warranties on pollution control equipment, including the catalytic converter.

Another remedy for emissions is "clunker" legislation, which pays people to voluntarily remove old vehicles from the road—older vehicles, especially those built before 1980, cause far more pollution than new ones. The oldest 10 percent of the cars on the road cause 50 percent of the hydrocarbon and carbon monoxide emissions.

Due to these enormous strides in reducing automotive emissions, today's cars are 90 percent cleaner than those of the early 1970s (United Nations, 2002). It would seem that with these significant technological improvements, the problem of vehicular pollution has been solved. Unfortunately, you only have to see the smog over large cities on hot summer days to realize this is not the case. What has happened? Two very simple things: first, the population continues to grow; and second, the standard of living has also increased. The result is more people owning more cars and driving them further.

The net result is that, in spite of enormous improvements in pollution control, total emissions from vehicles are greater today than they were in the 1970s. Gasoline- and diesel-powered vehicles are a major source of air pollution, and across Canada contribute 70 percent of carbon monoxide, 50 percent of NOx, 30 percent of volatile organic compounds (VOCs), 25 percent of carbon dioxide, and 65 percent of benzene emissions (Government Working Group, 1998). In cities, vehicle emissions contribute even more pollutants.

Many argue that technology will solve environmental problems. But in spite of strong technological innovation, this is not the case for automobile emissions.

At the same time that car use has grown, urban transit rides per person per year have declined to less than 40 percent of what they were four decades ago. Cheap parking, subsidies for highway construction, low fuel prices, and other hidden subsidies encourage car dependency and promote urban sprawl. And so the air in big cities continues to degrade.

In the United States, fuel-efficiency standards for new cars and light trucks are set under the Corporate Average Fuel Efficiency Standards, established in 1975. In Canada, fuel-efficiency standards were never enacted and instead voluntary agreements are reached with the car manufacturers to meet standards equivalent to those in the United States. These standards led to a 60 percent improvement in average fuel efficiency of automobiles by about 1987 (United Nations, 2002), which represents an enormous saving in oil.

SUV INSANITY

In 2003, fuel-efficiency standards in the United States were set at 8.6 L/100 km (27.4 mi. per US gallon) for new cars and 11.4 L/100 km (20.6 mi. per US gallon) for new light-duty trucks (pickups) and SUVs. That SUVs—the fastest growing segment of car sales until 2008—should be allowed to create 33 percent more pollution than regular automobiles seems incomprehensible. Now high fuel prices are doing the job that government should have done years ago.

The trend toward larger vehicles has been happening at the same time that the average size of North American families is decreasing. When asked why they own a large SUV when they never do any off-road driving, most owners respond that they think their children will be safer in an SUV than in a smaller vehicle. Unfortunately, this is a mistaken belief, for SUVs are not safe, either for the people who drive inside them (they topple more easily than regular cars), or for the smaller cars that share the road with them. And they are especially not safe for the environment. SUV owners' concern for their children's safety apparently does not extend to health, nor to the welfare of their grandchildren. The gas-guzzling SUV is a monument to North America's high standard of living and its lack of concern for the environment.

Since 1987, however, the fuel efficiency of new cars has steadily declined. In fact, because of the trend toward SUVs and pickup trucks, the overall fuel efficiency of new cars and light trucks actually worsened by about 13 percent between 1987 and 2007 (Pew, 2007), with an accompanying increase in greenhouse gas emissions.

Diesel Engines

Diesel engines, invented in 1890 by the German engineer Rudolf Diesel, are the workhorses of modern transportation and heavy industry. They operate by compressing air to a temperature sufficiently high to ignite fuel injected into the cylinder. Diesels do not need the ignition apparatus—spark plugs, rotors, coils, distributors, magnetos—of gasoline engines. Not only is the engine simple and dependable, but it develops higher power than a gasoline engine does at low speed, exactly when it is most needed. Diesel engines are preferred for heavy-use vehicles such as long-distance trucks, buses, and industrial equipment, because they are extremely robust and the most efficient of all internal combustion engines. The use of diesel engines will increase in the future. For example, truck traffic is predicted to grow twice as fast as rail or marine shipping (Environment Canada, 2001).

Diesel engines, however, are major contributors to air pollution, emitting much higher levels of contaminants than gasoline engines. Diesel engines burn fuel at low temperatures, resulting in incomplete combustion. The problem is compounded by the high concentrations of sulfur and other impurities in diesel fuel, which is not as highly refined as gasoline. Although advances have been made in improving diesel-engine efficiency and reducing emissions, trucks and buses are not subject to the same rigorous regulatory control as cars. The major pollutants from diesel trucks are particulates, sulfur oxides, and NOx, whereas VOCs and carbon monoxide are relatively minor. In Canada, estimates showed that in 1995 diesel trucks produced 30,700 tonnes (33,800 tons) of $PM_{2.5}$, or 37 percent of the total produced by the transportation sector (Environment Canada, 2001).

Since 1998, sulfur content in on-road diesel fuel was set at 500 ppm. There were no restrictions on off-road diesel fuel, which includes use in trains, boats, and machines and accounts for about 45 percent of the total diesel transportation consumption.

In the United States, the EPA requires serious reduction of diesel pollution from new heavy-duty diesel trucks and buses. For example, it mandated that sulfur content in diesel fuel be reduced from the allowable content of 500 ppm in 2002 to 15 ppm by 2006.

With the growing recognition that particulate matter is one of the deadliest air pollutants, regulations are changing. In 2006, Canadian regulations stipulated that sulfur content in on-road diesel fuels be lowered to 15 ppm from the previous 500 ppm. Starting in 2007, the maximum sulfur content for off-road diesel fuels could not exceed 500 ppm (previously there was no limit), and in 2010 will drop to 15 ppm. For rail and marine diesel fuels, a maximum sulfur content of 500 ppm is applicable starting in 2007, and this will drop to 15 ppm in 2012 (Environment Canada, 2006a).

These sulfur reductions should result in a direct reduction of particulate matter in the atmosphere. This was demonstrated in Denmark, where the sulfur content of fuels was reduced to 50 ppm in 1999. Peak particulate levels measured on the streets of Copenhagen dropped by more than 50 percent, with a significant shift in size distribution from small particulates to larger, less health-damaging particulates (Environment Canada, 2001).

The EPA is making major efforts to reduce emissions from trucks and buses. In 2007, the standard for NOx emissions was a maximum of 0.2 grams per brake horsepower hour (compared to 4 grams in 1998), and PM emissions must not exceed a maximum of 0.01 grams per brake horsepower hour (compared to 0.10 grams in 1998) (EPA, 2002). Canadian regulations match the US standards. (Note that because truck loads vary so widely, it is meaningless to speak of emission efficiency in terms of grams of emission per volume of fuel or per distance traveled. Instead, emission efficiency is measured in units of work done. Since brake horsepower is a common way of measuring the power of a truck, the weight of emission per brake horsepower hour has become the common way of expressing a truck's emission efficiency. This allows small and large diesel engines to be compared.)

The Pain of Gridlock

Automobiles extract a toll not only from their emissions but also through their vast numbers. Gridlock has become a major problem that takes a big bite of time out of commuters' personal lives, reduces productivity, causes enormous stress including road rage, and has been implicated in physical ailments including high blood pressure, lower back pain, and heart disease. No one enjoys sitting for hours inching forward on a fume-filled highway with the temperature of the engine, not to mention that of the occupants, slowly rising. Yet these experiences are all too common, and avoiding gridlock has become a major factor in day-to-day traveling and choosing a job or a place to live. It is an important political issue in large cities.

In 2006, a Canadian study estimated that gridlock in Canada's nine largest cities costs more than Can$3.7 billion each year (Taylor, 2006). This figure includes work time lost (calculated at $28 per hour), the cost of wasted gasoline, and the cost of greenhouse gases emitted.

A similar study conducted in the United States for 85 cities placed the cost of congestion at $63 billion. Between 1982 and 2005, the time wasted by the average rush-hour traveler has increased from 16 hours to 47 hours per year. That is, the typical American commuter wastes more than a full workweek each year.

Los Angeles is renowned for its traffic congestion. In 1987, it became the US city with the worst traffic, and has held that title ever since. Each year traffic has grown worse, to the point where it is now the single biggest factor preventing employers from attracting and retaining employees. Traffic reduces the willingness of Angelenos to attend social, cultural, and sports events and to visit friends and relatives. The average resident spends about 368 hours per year—15 full days—commuting to and from work (Diamond, 2005).

LONDON RESTRICTS CAR ACCESS TO CITY CENTER

In a bold move to relieve crippling traffic congestion in the center of London, England, drivers are being charged about $12 a day to enter a central area of 20 km² (7.7 sq. mi.) during business hours. The license numbers of vehicles in the controlled zone are recorded by about 700 surveillance cameras and fed into a computer database. Drivers who have not paid the toll by midnight are issued fines that escalate up to about $300 if payment is late. The program, which began in February 2003, is being watched closely by other major cities throughout the world. After one week, the predictions of chaos and mass civil disobedience failed to materialize, and the program appeared to be working smoothly. In fact, it was reported that traffic was moving so freely that buses had to slow down to keep to their schedules (Freeman and Lewington, 2003).

Other cities around the world are moving towards similar models of curtailing the traffic in their cores. Business-day access to the center of Athens, a city renowned for its thick smog, is restricted to vehicles with even-numbered license plates on even-numbered days and to those with odd-numbered licenses on odd-numbered days (European Environment Agency, 1999).

BICYCLES ARE GAINING POPULARITY

In the summer of 2007, Paris introduced an ambitious program to get Parisians out of their cars and onto bicycles. More than 10,000 bicycles were placed at 750 stations throughout the city. For a small fee riders can take bikes from one station and drop them at another. The

"Velib" (short for the French phrase for "free bicycle") program proved so popular, the number of bikes was doubled in 2008. Several other European cities are introducing similar programs.

Transportation in Context

The bottom line is that, in spite of immense reductions in automotive emissions, air quality has deteriorated. It is clear that to reduce transportation-caused air pollution, much more is required. Good progress has been made with cars and trucks running more cleanly and efficiently. Now the same standards must be applied to off-road vehicles, trains, ships, and planes. But as history has shown, this will most likely be insufficient. The number of vehicles, the total distances traveled, or both must also be curtailed. Given an ever-growing population and increasing standard of living, this is a Herculean task and will require changes in the ways we live and do business.

A major obstacle to reducing air pollution from the transportation sector is the sheer number of cars and trucks on the road. And these numbers continue to increase. A good example of this is California, as seen in the table below which shows the enormous number of people and vehicles in 2000, with projections to 2050 (CNGVP, 2004).

TABLE 5-1: Estimates of Population and Number of Cars in California for 2000 and 2050

	2000	2050
Population	33.8 million	67 million
No. of cars	24 million	47 million
Distance driven — km (mi.)	480 (300) billion	871 (541) billion
Fuel consumed — liters (gallons)	64 (17) billion	140 (37) billion

Programs should be implemented that reduce the *total* distance traveled (not the *per capita* distance) and at the same time decrease the emissions per unit distance traveled. For example:

• Regulations should continue to set progressively more restrictive emission standards.

- Regulations should continue to set progressively more restrictive fuel-efficiency requirements.
- Higher taxes should be imposed on vehicles such as SUVs that are less efficient and emit higher amounts of pollutants.

But these actions are treating only the symptoms of the problem. We also need to get at the root problem and decrease our dependency on the motor vehicle in a way that preserves or enhances our quality and enjoyment of life. We need to rethink many of the principles that we have held sacred for decades.

In North America, major cities are characterized by sprawling suburbs with row upon row of large single-family homes far removed from the city center—the so-called American dream. This sprawl not only makes the automobile vitally necessary, but also leads to significant social, economic, and environmental costs. We need to shift to communities that are more compact and friendlier to mass transit, pedestrians, and bicycles.

THE HIGH COST OF URBAN SPRAWL

Farmland is the first casualty of uncontrolled urban sprawl. Between 1982 and 1992, an average of 1,620 km² (626 sq. mi.) per year of prime farmland in the United States was urbanized (United Nations, 2002). During the following decade, this loss rate doubled as almost 65,000 km² (25,100 sq. mi.) of private forest, agricultural land, and open spaces were claimed by development, so that by the end of the 1990s, some 4,050 km² (1,560 sq. mi.) were lost each year. In Canada, from 1971 to 1996 approximately 12,250 km² (4,730 sq. mi.) of land—half of which was dependable agricultural land—was converted to urban sprawl. With world grain production falling short of consumption in seven of the last eight years, and now with large swaths of productive land producing plants for fuel ethanol instead of food, there is an urgent need to increase, not decrease, farmland.

Sprawl is also a serious threat to wildlife and plants, as it destroys or degrades the habitats on which they depend for their survival.

We need to move toward "smart growth" and "sustainable cities", which encourage the development of high-density neighborhoods characterized by a balance of mixed residential, office, and retail uses close to each other and to civic buildings. Here, large cities effectively become a grouping of neighborhoods that resemble small self-sufficient towns. In these group-ings, travel distances shrink, walking and cycling are encouraged, and pub-lic transit is economically viable. Smart growth also preserves open green spaces and farmland, and includes revitalization of city cores and redevel-opment of brownfields (previously contaminated industrial sites).

SMART GROWTH IN PORTLAND

In the 1970s, city leaders in Portland, Oregon, frustrated by urban sprawl and gridlock, decided to promote public transit and sensible development by implementing careful land-use restrictions to growth. Portland became the most progressive community in the country and was watched closely as a model for smart growth. Some facts:

- Although Portland's population has grown by 50 percent since 1973, its land area has only grown by 2 percent.
- In spite of the population growth, with its attendant increase in the number of cars, air quality consistently improved from 1990 to 1999 with bad air days decreasing from 11 to only 2 per year.
- Transit ridership increased 46 percent from 1996 to 2006, while auto-mobile miles driven per capita dropped about 8 percent.

Portland's move to smart growth has also led to more cycling, re-development of inner-core industrial areas, "green" buildings, and a pride in recycling. Portland regularly ranks at the top of America's greenest cities. In 2008, for example, Portland was selected—for the fourth year in a row—as the most sustainable of America's 50 most populous cities by SustainLane using 16 indicators such as public tran-sit, renewable energy, local food, and development approaches.

Many regulatory steps can and should be implemented to encourage smart growth. These include the following: reducing taxes on buildings and increasing taxes on land; eliminating tax subsidies for free employee parking and making subsidized transit passes tax deductible; introducing road tolls and increasing fuel taxes; instituting pay-per-distance car insur-ance; and improving and subsidizing transit services.

The vested interests of the car manufacturing industry are so powerful, and suburban sprawl is so entrenched in the North American lifestyle, that reversing the trend presents a formidable challenge.

GASOLINE TAXES

Even though a large part of the cost of gasoline consists of taxes, it was shown in Vancouver, British Columbia, that these taxes, even when combined with license fees, cover only about 80 percent of the costs of building and maintaining roads. Furthermore, the health-care costs resulting from road accidents and vehicular air pollution are not covered by these taxes (Tollefson et al., 2000). In other words, car drivers are subsidized by the government to drive and pollute the air. Perhaps it is time to increase the price of gasoline—as has been done in Europe—to reflect its true cost.

Although significantly increased use of mass transit would relieve traffic congestion and improve air quality, it is not a popular mode of travel in North America. The total distance driven by automobiles in Canada has increased more than fourfold since 1950 and accounts for over 95 percent of travel. Over the same period, urban transit ridership has remained about constant, accounting for less than 5 percent of the passenger distance traveled. Mass transit needs to be made more appealing and convenient, for example, by providing free parking at outlying stations, television screens in subways, and improved access to transit schedules.

The widespread use of personal computers linked by the Internet and the World Wide Web has made it possible for many people to work from home. This leads to continuing growth in home-based businesses and also allows employees of large firms to remain in close touch with fellow office staff while working at home. Likewise, teleconferencing, often aided by electronic imaging, is becoming increasingly common.

The railway is to freight transportation what mass transit is to people transportation. The use of railways should be increased significantly, for this would not only improve the environment and relieve congestion, but would also avoid considerable carnage through road accidents.

CARS AND TRUCKS GO IN OPPOSITE DIRECTIONS

American roads and highways are the scene of a significant conflict. On the one hand cars need to become smaller and more fuel efficient. On the other hand, trucks are getting larger, with the biggest trucks pulling three trailers that can be as long as a ten-storey building is tall. Behemoths like this should not be sharing the road with cars; the trend is certain to increase the highway death toll. In 1999 in the United States, more than 400,000 trucks were involved in traffic accidents, with 5,362 people killed and an estimated 142,000 injured. Fatality rates for accidents involving large trucks are about 65 percent higher than for those involving only passenger vehicles.

Greener Cars

Ultimately, we need to shift away from our dependency on fossil fuels by developing alternative fuels and advanced cars that emit far less pollution. This is a challenging task, which will take at least several decades to implement. A number of alternative fuels have been developed with varying degrees of success. The main ones are propane, natural gas, and ethanol. Of the total number of alternative-fuel cars in use in the United States in 2002, the liquefied petroleum gases (propane) and compressed natural gas dominate the market.

Alternative Fuels

Propane or liquefied petroleum gas (LPG) has been used as a transportation fuel for more than 60 years, and is the alternative fuel most widely used at this time. LPG consists mainly of propane, propylene, butane, and butylene in various mixtures. (In North America the mixture is mainly propane.) It is a byproduct of natural gas processing and petroleum refining, and is popular because an infrastructure of pipelines, processing facilities, and storage already exists for its distribution. Almost 300,000 light- and medium-duty vehicles, mostly in company fleets, are traveling American highways under propane power. According to the National Propane Gas Association, spark plugs from a propane vehicle last from 129,000–161,000 km (80,000–100,000 mi.) and propane engines can last two to three times longer than gasoline or diesel engines. Like natural gas, propane contains much less sulfur than gasoline, so it emits less sulfur dioxide and fewer particulates.

Natural gas is increasingly popular as an alternative fuel because it is cheaper than gasoline and produces significantly fewer harmful emissions. Natural gas can be stored onboard a vehicle in a tank, either as compressed gas or cooled to such low temperatures that it forms a liquid. Natural gas is distributed throughout North America in extensive pipeline systems, which extend from the wellhead to the end user. Currently, natural gas is the second most common alternative fuel.

Ethanol (ethyl alcohol or grain alcohol) has become a hot issue and is leading the charge in the rush to biofuels. It is a clear, colorless liquid with a characteristic agreeable odor, made from crops that contain sugar, such as sugar beets and sugar cane, or from grain crops that contain starch or cellulose that can be converted into sugar, such as corn. The US Clean Air Act Amendments of 1990 mandated the sale of oxygenated fuels in areas with unhealthy levels of carbon monoxide; this led to a growing demand for ethanol. In 2004, the Canadian and Ontario governments initiated steps to ensure that much of the gasoline supply would contain 10 percent ethanol by 2010. In the United States, more than 1.5 billion gallons of ethanol are added to gasoline each year to increase octane and improve the quality of gasoline emissions. Ethanol is usually blended with gasoline to form an "E10" blend (10 percent ethanol and 90 percent gasoline) sometimes called "gasohol", but it can be used in higher concentrations (85 percent ethanol is common) or in its pure form. Ethanol blends are successfully used in all types of vehicles and engines that require gasoline. Another blend is "E-diesel", a mix of up to 15 percent ethanol and regular Number 2 diesel fuel; this is starting to be used in diesel vehicles without any major engine modifications.

The great hope for ethanol is that it will provide "oil security" while reducing air pollution and global warming. For example, it emits 40–80 percent less greenhouse gases than regular gasoline does. Furthermore, ethanol is made from renewable resources.

Critics argue, however, that planting, growing, harvesting, crushing, fermenting, and distilling corn to make ethanol uses large amounts of fertilizer and requires more energy than the ethanol produces when it is burned. This leads to environmental damage, a net increase in total energy consumed, and most likely a net increase in greenhouse gases created (Mihlar, 2003). The American Lung Association claims that ethanol results in higher emissions of nitrous oxides and it evaporates more easily in the event of spills.

Proponents argue that deriving ethanol from corn can be made more efficient. One way, for example, would be to develop new enzymes that would provide a small net energy gain. The ultimate goal is to develop cellulosic ethanol, which is made from the cellulose in plants; this would mean that not only the corn kernel but the entire plant could be used. Other plants that are native to the prairies, which would need little or no fertilizer, could also be used. Cellulosic ethanol cannot yet be manufactured commercially, but intensive research is underway (Sanford, 2006). The promise of ethanol has been realized in Brazil, where it is made from sugar cane, which is twice as productive in making ethanol as corn. After the OPEC oil crisis in the 1970s, Brazil began a program to develop ethanol fuel quickly, and today about one-third of the fuel Brazilians use in their vehicles is ethanol. This compares to about 3 percent in the United States (Morgan, 2005).

A big problem with ethanol is that it contributed to a dramatic increase in the price of wheat and other grains. It is an ominous sign when vehicle fuel and bread are locked in head-to-head competition, with the prices of both of these basic commodities skyrocketing. In 2007, about 73 million tonnes (81 million tons) of US corn were used to produce ethanol. This represented 20 percent of the country's grain crop, but the resulting ethanol was less than 4 percent of its automotive fuel (Brown, 2008).

Biodiesel (fatty acid alkyl esters) is a diesel replacement fuel made from new and used vegetable oils and animal fats (DOE, 2008). Blends of up to 20 percent biodiesel (mixed with regular diesel fuel) do not require any engine modifications and can provide the same payload capacity and range as diesel. Higher blends and even pure biodiesel ("B100" or 100 percent biodiesel) can be used in many engines built since 1994 with little or no modification. Using biodiesel in a conventional diesel engine substantially reduces emissions of unburned hydrocarbons, carbon monoxide, sulfates, polycyclic aromatic hydrocarbons, nitrated polycyclic aromatic hydrocarbons, and particulate matter. Because biodiesel contains less than 24 ppm sulfur, it works well with new technologies such as catalytic converters, particulate traps, and exhaust-gas recirculation. Of the vegetable oils, the least expensive is soy oil—and it is no surprise that the soy industry has been the driving force behind biodiesel commercialization. The use of biodiesel has grown dramatically during the last few years and includes the fleets of the US Postal Service and the US Departments of Energy and Agriculture.

Hydrogen gas (H_2) has the potential to play an important role in sustainable transportation in the future because, in theory, it can be produced in virtually unlimited quantities and it creates no harmful emissions.

Hydrogen would be used in fuel-cell vehicles in a process that is the reverse of electrolysis (separating water molecules into hydrogen and oxygen with an electric current). Hydrogen and oxygen would be combined to create electricity, which then powers the vehicle. In other words, although the fuel is hydrogen, the car is driven by electricity and the only emission is water. Internal-combustion engines in modern cars convert less than 30 percent of the gasoline's energy into power to move the vehicle. In comparison, vehicles using electric motors powered by hydrogen fuel cells use 40–60 percent of the fuel's energy.

The making of hydrogen requires electricity, and unless the source of electricity is "clean", the use of hydrogen vehicles merely removes the source of pollution from the vehicle to the power plant. At normal temperatures hydrogen is a gas, which poses storage problems. Storage methods under development include compressed hydrogen, liquid hydrogen, and chemical bonding between hydrogen and a storage material (for example, metal hydrides). Pure hydrogen and hydrogen mixed with natural gas (for example, Hythane) have both been used effectively to power cars.

The quest for a "hydrogen highway" has been difficult. Although major car companies are investing in research, progress is slow. DaimlerChrysler is one of the leaders; since beginning hydrogen experiments in 1994, it has built more than 100 fuel-cell vehicles. A fleet of 30 buses has been carrying passengers daily in ten European cities, and 60 Mercedes-Benz A-Class sedans converted from gasoline engines to fuel cells have logged 3.2 million km (2 million mi.).

An ambitious "hydrogen highway" is planned for the Vancouver Olympic Games in 2010, stretching from the Vancouver airport to the Whistler ski resort. Intended to stimulate research into and commercialization of hydrogen fuel cells, it will include fuel-cell buses and cars, and hydrogen fueling stations.

Electric Cars

Electric cars powered by batteries offer a clean choice of transportation, because the vehicle itself emits no pollutants at all. The pollution is removed to the location of the power plants that supply the electricity.

An electric car, the Honda EV+ had a range of about 160–200 km (99–124 mi.) in 1998; this makes it viable for city usage, but not for long trips. Research is underway to develop batteries of many different compositions. Improvements in range will undoubtedly be seen in the future. For example, the second generation General Motors EV1 electric car that was unveiled in 1999 features a nickel metal-hydride battery with a range reaching over 210 km (130 mi.). Additional benefits of electric cars include the following: their fuel—electricity—is cheaper than gasoline; and they require less maintenance and last longer than regular cars because they have fewer moving parts. The acceptance of electric cars is greatest in Los Angeles County, where in 2006 there were 112 electric-vehicle charging stations. Public charging stations are springing up in major cities throughout the world as the popularity of electric cars grows. Because batteries are heavy and have limited range, these cars have not yet significantly penetrated the market. But research continues.

Hybrid Cars

A new type of "green" car emerged in 2000, when Toyota launched the Prius, and Honda rolled out the Insight and the Civic Hybrid. These hybrid cars combine an electric battery with an internal combustion engine to reduce emissions and increase gas mileage. The Honda Insight, ranked number one in fuel efficiency, burns a mere 3.44 liters of gasoline for every 100 km (68.4 mi. per US gallon). The Insight is powered by a 1.0 liter, 3 cylinder gasoline engine that receives additional power from the battery when needed; for example, while accelerating from a stop. In turn, the battery is charged from the energy dissipated when the car is braking—a clever use of energy that would otherwise be wasted.

Because the hybrid-car sticker prices are about $3,000 to $4,000 higher than a comparable conventional car, sales were slow to take off. But with the huge increase in gasoline prices in 2006, the popularity of hybrids skyrocketed. Now other car companies are rushing to bring out their hybrid models, including hybrid SUVs. In 2006, there were 1,234,655 alternative fueled and hybrid vehicles sold in the United States (EIA, 2008a). With oil over $120 per barrel, hybrid sales have spiked upward and the hybrid car is a commercial success.

The "plug-in hybrid" is like a regular hybrid but with an extension cord. With a bigger battery that can be plugged in at home to recharge from the electrical grid, the hybrid is in effect converted into an electric car with a small backup gasoline engine. Short distances can be traveled almost entirely on electricity, saving expensive gasoline for longer journeys. In 2008, General Motors announced its intention to mass-produce two plug-in hybrids, a Saturn VUE SUV, and the Chevy Volt, which should be available to the public in 2010. Toyota has announced it will have a commercial plug-in hybrid on the road by 2010.

CALIFORNIA AND CLEAN-AIR CARS

Spurred by the enormous smog problem in Los Angeles, California has traditionally been a leader in North America in developing air pollution regulations. In the early 1990s, the California Air Resources Board (CARB) passed legislation requiring 5 percent of all vehicles sold in the state in 2001 be "zero-emission vehicles", and that this would increase to 10 percent (about 100,000 vehicles) by 2003. A percentage of California vehicles would also have to be "ultra-low emissions vehicles" using one or both of super-clean gasoline engines or alternative fuels such as propane, alcohol, and natural gas. In 2003, in response to intense lobbying by the car industry, California decided to give the auto industry more flexibility in introducing clean-air technology. The emphasis is on developing plug-in hybrids, hydrogen fuel-cell vehicles, and vehicles powered by natural gas. A large increase in the number of plug-in hybrids is expected from 2009 to 2015.

A Complicated Issue

This chapter highlights a frustrating problem. The automobile is a wonderful invention that greatly enhances our productivity and enjoyment of life. On the other hand, the internal combustion engine emits harmful pollutants and burns a valuable resource. Over the past three decades some of the brightest minds in the world have worked to overcome these problems by making cars cleaner and more efficient, with enormous progress. Catalytic converters and positive crankcase ventilation have been introduced; lead has been removed from gasoline, and sulfur greatly reduced in fuel; vehicle fuel efficiency has been improved; and the list goes on.

But in spite of all this development, there has been no meaningful improvement in air quality. All of the world's major cities continue to gasp under poisonous domes of smog. The reason is simple: there are ever more cars on the road and they are being driven ever farther.

The prospect we face is simple, yet frightening. The growing population and expanding economy cancel out technical advances in pollution prevention. It is like being on a treadmill—the population-economy treadmill—whose speed is constantly being increased; no matter how fast we run, we can't make any progress.

Perhaps a different approach is required. Perhaps we need to think seriously about stabilizing population levels and changing our lifestyles to place less emphasis on consumerism. We revisit this topic in Chapter 11.

6

Acid Rain: The Scourge of the Loons

It was an annual ritual, as sacred as Christmas or Easter. When the hot, lazy dog days of summer arrived, my parents would pack the car and bundle my brothers and me into the back seat. With the windows wound down to catch the cooling breeze, we headed to our cottage deep in the forests of northern Ontario. Our favorite activity was canoeing; we spent many hours every day, like junior voyageurs, exploring the network of lakes and the primeval landscape of rocky outcrops and thick bush.

In those days loons were plentiful, and whenever we saw one swimming nearby we would give chase, paddling like maniacs. But every time we got close, the loon dove underwater. My brothers and I made bets on where it would emerge and we tried desperately to hold our breath as long as the loon did. We always gasped for air long before the loon appeared and never correctly picked the spot where it popped up, for the loon is a clever bird and can swim long distances under water. Later, at dusk around the campfire, the cry of the loon would occasionally break the stillness— a haunting ululating sound that hushed our conversation.

When I recall my childhood, I invariably think of canoeing and loons. Sadly, when my children were young and paddled the same lakes, they rarely saw loons. Times had changed, and the industrial might of society had extended its deadly reach far into the forests and lakes of northern Ontario. Air currents brought to this region something ominous and invisible that was a scourge for the loons and many other forms of life. It was acid rain.

Acid rain—or "acid deposition" as it is called by scientists—is a complex problem that is both regional and international in scope. It has pitted one region of the United States against another, strained relations between Canada and the United States, and caused bitter acrimony among European countries. Over the past three decades, however, the problem of acid rain has been recognized and understood. In the developed nations, if acid rain has not been entirely stopped, it has at least been reduced. Measures are in place to continue the fight against it. In developing nations, however—especially in China and India—acid rain is a major and growing environmental threat.

Acid rain is caused by power plants, factories, vehicles, and homes that burn fossil fuels containing sulfur as an impurity, which is released into the atmosphere as sulfur dioxide (SO_2). Processing sulfur-bearing ores also emits sulfur dioxide. Nitrogen oxides (NOx)—sulfur's "partners in crime"—are generated during high-temperature combustion with nitrogen coming from the air or impurities in the fuel. The sulfur and NOx join with moisture in the atmosphere to form sulfuric and nitric acids, which subsequently fall to the ground. This causes both health problems and damage to the environment.

History

The existence of acid rain and the damage it causes to the natural surroundings were known long ago. In 1872 the British chemist Robert Angus Smith, in his book *Air and Rain: The Beginnings of a Chemical Climatology*, wrote, "Acidity is caused almost entirely by sulphuric acid, which may come from coal or the oxidation of sulphur compounds ... The presence of free sulphuric acid in the air sufficiently explains the fading of colours in prints and ... the rusting of metals, and the rotting of blinds" (Pringle, 1988). Smith's observations were ignored by society for almost 100 years, but became quite relevant in the latter half of the twentieth century.

The annual emissions of sulfur dioxide in Europe were relatively constant at about 25 million tonnes (28 million tons) from 1910 to 1950. Then they rapidly tripled, so that by 1980 emissions had reached 75 million tonnes (83 million tons) per year (Gribbin, 1986). In 1961, Svante Odén, a Swedish scientist, first recognized that acid rain had become a serious environmental problem in Scandinavia. After a few years of collecting water samples, he noticed a correlation between acidity in lakes and that of rain from the European sampling network. In 1967, he observed fish dying in lakes that had become highly acidic. Similar problems had been observed by others for decades. For example, dying Atlantic salmon and brown trout had long been noted, but Odén was the first to link the problems to air pollution. He published his results in 1968, and the public finally began to come aware of this new environmental threat.

Odén's studies of acid rain marked an important milestone: previously, pollution problems had always been thought to be local. Now the scale of environmental degradation had expanded greatly, and there was an urgent need to understand how pollutants interacted with and were carried by regional air flows and even global wind patterns. Suddenly, it seemed that no area in the world, no matter how pristine and free of industry, was safe.

The usually mild-mannered citizens of Sweden were outraged when it was revealed that about 77 percent of Sweden's acid rain came from other countries—industrial smokestacks in Great Britain and Germany's Ruhr Valley—which they saw as "a form of unpremeditated chemical warfare" (Pringle, 1988). From roughly 1956 to 1976, the acidity of rain in southern Norway and southwestern Sweden increased by 200 percent. In 1980, data suggested that about 20 percent of Sweden's lakes were affected (Gribbin, 1986). By 1983, fish life had been destroyed in 5,000 lakes in Sweden and 1,500 in Norway. Salmon and trout were the first to die and restocking was not successful (Young, 1990). Ominously, Sweden warned that acid rain could also be a problem in the northeastern United States and eastern Canada.

Acid rain was very much in the public eye in the late 1970s and throughout the 1980s, when areas of central Europe suffered extreme degradation. Bavaria was hardest hit with up to 40 percent of trees sick or dying. The Germans called this phenomenon *Waldsterben,* or "forest death", a chilling name that captured media headlines. Studies tracing the origins of this threat soon revealed that sulfate concentrations were greatest in densely populated areas, and especially near (particularly downwind from) factories that burned coal as a fuel. The acidity of lakes was greatest downwind of industrial areas.

In the mid-1970s, Canada and the United States set up a long-term monitoring program; not surprisingly, it was soon reported that acid rain was also causing serious damage in North America. Acid rain, in fact, turned out to be a particularly Canadian problem, with many thousands of northern lakes so acidic that they were essentially dead. The high-elevation spruce forests of the northeastern United States, as well as the deciduous forests of eastern Canada and the United States, had also come under attack.

Many studies followed: the biggest in the United States was the National Acid Precipitation Assessment Program (NAPAP). This program began in 1980, involved 700 scientists, cost half a billion dollars, and spanned a decade. The results were variable, demonstrating that acid rain in the United States was a serious problem, although not quite as serious as had been first speculated. In Canada, however, the situation was worse than previously thought.

The public was incensed and, consequently, strong regulations and measures were put into place to control sulfur dioxide emissions. In particular, the use of lower-sulfur coal and the installation of pollution control equipment such as scrubbers and electrostatic precipitators proved effective (see Chapter 7). Things began to improve.

The situation also improved in Europe. In the late 1990s, the United Nations reviewed the state of the world's forests and concluded: "the widespread death of European forests due to air pollution which was predicted by many in the 1980s did not occur". The problem in Bavaria was caused by emissions from local industry, which were subsequently regulated. When sulfur dioxide went down 50–70 percent in seven years, forests generally recovered (Lomborg, 2001). The situation, although improved, is not completely resolved. In 2002, about 21 percent of all German forests displayed visible defoliation, and some effects of acid rain are only recently beginning to appear. This is particularly the case on the forest floor, where the soil is acidifying after absorbing pollutants for decades (EIA, 2006).

By contrast, in China, India, and other developing countries where large quantities of coal are burned to generate electricity—often with little or no pollution control—acid rain has become a very serious problem. In fact, sulfur dioxide is China's number one pollution problem, causing serious health and environmental damage (Bradsher and Barbosa, 2006). In 2004 China released about 20.4 million tonnes (22.5 million tons) of sulfur, more than twice the amount released in the United States. This amount will probably increase substantially in the coming years with the rapid construction of new coal-fired power plants.

Acid-Rain Chemistry

The term "acid rain" seems relatively straightforward; however, the chemistry behind it is actually very complex. For example, sometimes acidic compounds do not form until after airborne sulfates and nitrates come into contact with water on the ground. And these acid-causing chemicals can appear even in clear weather, not just during rain or other precipitation. For this reason, scientists prefer to call it "acid deposition" rather than acid rain.

Unpolluted rainwater is slightly acidic, with a pH of about 5.6; this is a result of the carbon dioxide that is naturally present in the air combining with the water to form a weak carbonic acid. Ice core samples from Greenland that are about 180 years old—and which predate sulfur emissions and the recent increases of atmospheric carbon dioxide—yield water that is less acidic, with a pH of 6–7.6 (Pringle, 1988).

HOW SCIENTISTS MEASURE ACIDITY

Chemists use the term "pH" to measure the acidity of fluids. The scale ranges from 1 to 14, where 1 is the most acidic, 7 is neutral, and 14 is the most basic or "alkaline". Acidity increases as values approach 1, and as they approach 14, alkalinity increases. The scale is logarithmic, so each integer denotes a tenfold increase or decrease. For example, a substance with a pH of 3 is ten times more acidic than one with a pH of 4.

Sulfur dioxide is a colorless gas that smells like burnt matches. In the atmosphere it oxidizes to sulfur trioxide (SO_3), which combines readily with water vapor to form sulfuric acid (H_2SO_4) in the form of mist or particles that can be transported long distances by prevailing wind currents.

In the atmosphere, NOx change to nitric acid (HNO_3) in hours, whereas the conversion of sulfur dioxide to sulfuric acid may take up to several days, taking place most quickly in summer sunlight. Both can fall to the ground as small particles, in solution in rain or snow, and even as gas that is taken up indirectly by plants. Acid in precipitation is easy to measure, but in dry deposition it is not. In southern California, because of its low precipitation, about 20 times more "dry" acid reaches the ground than does the "wet" form.

Rain at the start of a storm has the most acid as it flushes contaminants from the air. Similarly, the first snowmelt in spring, called a "freshet", is highly acidic. This can lead to temporary acid shock to lakes and rivers in early spring, which may be particularly lethal to fish and other aquatic life forms, even if a lake has "buffering capacity" (chemicals and minerals in its sediments and surrounding rocks that can neutralize acid) to recover fairly quickly. Unfortunately, acid shock often coincides with spawning time and can disrupt fish reproduction.

Sulfur dioxide accounts for about two-thirds of acid rain in the eastern United States and Canada. West of the Mississippi, NOx play a greater role, and in southern California NOx dominate due to the prevalence of pollution from cars and trucks (Pringle, 1988).

Sources of Acid Rain

Some acid rain comes from natural sources, mainly volcanoes, and also sea spray and decaying vegetation. For example, when it erupted in 1980, Mount St. Helens in Washington State emitted about 1,800 tonnes (1,980 tons) of sulfur dioxide per day, compared to 73,000 tonnes (80,500 tons) per day from cars (Cagin and Dray, 1993). Overall less than 5 percent of acid rain is caused by natural sources; the vast majority is from human activity.

Sulfur dioxide is produced by industrial processes and burning fossil fuels. Coal-fired power generation, ore smelting (given that many ores contain sulfur), and natural-gas processing are the main contributors. In 2000, US sulfur dioxide emissions were measured at 14.8 million tonnes (16.3 million tons), which represents a considerable reduction from the 24 million tonnes (26 million tons) per year in the mid-1980s. In 2000, Canada produced 2.4 million tonnes (2.6 million tons) of sulfur dioxide emissions, about one-sixth of US emissions.

The sources of sulfur dioxide emissions from the two countries are distinctly different. In 2000, in Canada, 68 percent of emissions came from industrial sources and 27 percent came from electric utilities. In the United States, electric utilities were the main contributor in 2002 at 67 percent of emissions (Environment Canada, 2006b).

For decades the largest single source of sulfur dioxide in North America was the Inco nickel and copper smelter in Sudbury, Ontario. Even after major investments in the installation of pollution control devices, the nickel smelters in Sudbury emitted about 0.3 million tonnes (0.33 million tons) of sulfur dioxide in 1999, over 43 percent of the total emissions in the province of Ontario (MOE, 2001). Air pollution—including sulfur dioxide from the sulfur-rich tar sands in Alberta—is growing rapidly.

The average sulfur content of crude oil in Canada in 1991 was 0.8 percent, although heavy fuel oil such as that used in electricity plants can contain up to 2.5 percent. Natural gas when first exhumed contains sulfur; the sulfur is virtually all removed during processing, although some escapes during exploration and processing. Alberta coal has the lowest sulfur in Canada at about 0.2 percent. Saskatchewan lignite coal is about 0.4 percent sulfur. In contrast, some coal burned in thermal plants in the Maritimes contained up to 6 percent sulfur (Standing Committee, 1993), although the last coal mine in the Maritimes closed in 2001. Ontario uses US coal at about 1–1.5 percent sulfur. In the United States in 2005 and in Canada in 2006, the standard for automotive gasoline was lowered to 30 parts per million (ppm) of sulfur.

High-sulfur coal, which contains more than 3 percent sulfur by weight, is found in northern West Virginia, western Kentucky, Pennsylvania, Indiana, Illinois, and Ohio. Low-sulfur coal, which contains less than 1 percent sulfur, is found in Virginia, Tennessee, eastern Kentucky, and southern West Virginia (Pringle, 1988).

In 1990, the worldwide generation of sulfur dioxide emissions was about 122 million tonnes (134 million tons) per year. At that time, the emissions from the United States and China were about the same, at about 20 million tonnes (22 million tons) per year. Since then the emissions of these two countries have taken different paths: China has increased its emissions, while those of the United States have decreased. The global emission of sulfur dioxide in 2005 had decreased to about 100 million tonnes (110 million tons), but this is projected to increase to 110 million tonnes (121 million tons) by 2030, mostly due to the expanding economies of Asian nations (acidrain.org, 2005). The economies of China and India are growing so quickly that future projections for sulfur dioxide emissions could easily be surpassed.

China, with its huge population and fast-growing economy, is a major contributor to acid rain and global warming. It is the world's largest user of coal, and many coal-fired industrial and electricity plants do not have even rudimentary pollution control. In 2005, coal- and oil-fired power stations were responsible for 11 million of the 25 million tonnes (12 million of 28 million tons) of China's sulfur dioxide emissions. It is not surprising that acid rain affects one-third of the country's territory (Clean Air Initiative, 2006).

Once airborne, acid rain may remain local, causing a sulfur smog; or winds can blow it a long way, making acid rain a regional and international problem. Scientists, using satellite imagery to track storms and weather patterns, have demonstrated that highly acidic rain in Massachusetts comes from the Ohio Valley and other midwestern states where coal-fired stations are concentrated. In the Rocky Mountain states, concern is focused on copper smelters in Arizona, Utah, Nevada, and New Mexico.

Acid rain recognizes no national boundaries. The acid haze over Alaska is thought to come from Japan. The acid rain that falls in eastern Canada comes from the US midwest. Canada claims that at least half of its acid rain is from the United States. In turn, the United States claims that 10–15 percent of the acid rain falling in its northeastern states is from Canada (Mitchell, 1991). China's pollution problem has become the world's problem. Sulfur dioxide and NOx spewed by China's coal-fired power plants fall as acid rain on Seoul, South Korea, and Tokyo, Japan (Kahn and Yardley, 2007).

Effects of Acid Rain

Natural ecosystems are extremely complex with numerous interactions at all levels, making it difficult and expensive to evaluate the effects of acid precipitation with precision. Acid rain affects soil chemistry, which in turn affects bacterial activity and plant growth. Acid rain also affects surface water; this interacts with soils. The acidity of both water and soil affects freshwater fauna and flora. Although all the fine details are not yet fully understood, many of the large-scale ecological effects are known.

Sulfur dioxide causes more damage than NOx because sulfuric acid is a stronger acid than nitric acid. Sulfuric acid is also considerably more destructive compared to the weak natural carbonic and organic acids that are buffered by calcium and magnesium in soils.

LOONS

In the late 1980s, Canadian biologist Robert Alvo studied 84 lakes near Sudbury, Ontario, which were some of the most acidic in North America given the large mining and smelting operations nearby. He observed a high death rate of newborn loons (although other birds including adult loons appeared to be healthy). He noted that 62 percent of young loons died on acidified lakes, but only 14 percent died on healthy lakes. The deaths generally occurred during the first 11 weeks following birth, before the baby loons could fly. Normally, loon hatchlings are fed for the first month on small insects, and after that on fish. Alvo watched as the parents dove repeatedly but could not find enough fish for themselves and their offspring in the acidified lakes. The parents would fend for themselves by flying to healthy lakes, but the hatchlings would die. For two years in a row, Alvo watched through his telescope and saw this disturbing scene unfold (Pringle, 1988).

Soil

When acid rain falls on the ground, the sulfuric acid forms a solution in the soil that mobilizes metals such as aluminum, cadmium, manganese, mercury, zinc, and copper. These metals then move through the soil, leaching out and replacing the useful plant nutrients such as calcium, magnesium, and potassium (Mitchell, 1991). Nitric acid does not have the same effect on soil metals, because nitrate is used as a nutrient by plants (Environment Canada, 1990).

Acid rain is a particularly Canadian problem: the eastern two-thirds of the country (the Canadian Shield) consists of hard granite, whose cover of soil was scraped bare by the glaciers of the last ice age. This region does not have alkaline rock such as limestone, and so cannot neutralize acid naturally. By contrast, in western Canada and southeastern Ontario, the geology consists of sedimentary rocks including limestone; water and soil systems are more alkaline and able to buffer against or neutralize acid rain naturally.

Lakes and Rivers

Sulfuric acid increases the acidity of lakes and rivers, damaging the flora and fauna that live there. The effects on lakes are much better understood now thanks to David Schindler of the Freshwater Institute of Winnipeg. Starting in 1976, he added sulfuric acid annually to a 27 hectare (68 acre) lake and studied the effects. A healthy lake has a pH of 7 or a bit higher. At pH 6 he noted that brook and rainbow trout populations started to decline and several species of clam and snail were wiped out. At pH 5, all but one species of crayfish were dead, as well as all brook trout, bullfrogs, and walleyed pike. At pH 4, all aquatic plants were dead or in decline except a few that could tolerate acid. At pH 2, the lake was sterile, with no living things at all. As acidity increased, the lake water became clearer and there was a shift from aquatic plants toward mosses and algae (Pringle, 1988, Mitchell 1991).

The impact of acid rain on a lake depends on the lake's ability to neutralize acid with naturally occurring alkaline substances such as limestone. This, in turn, depends on the surrounding soils, the bedrock below the lake, or both. Because the Canadian Shield has little buffering capacity, it is estimated that 142,000 lakes in eastern Canada and some rivers in Nova Scotia have been severely affected by acid rain since the 1970s (Environment Canada, 1990). In 2000, it was concluded that in spite of significant progress in decreasing sulfur dioxide emissions, acidity is still increasing in 11 percent of lakes in eastern Canada (Tollefson et al., 2000).

Plants

Acid rain damages the surfaces of tree leaves and needles, reduces a tree's ability to withstand cold, and inhibits the germination of pollen. Consequently, a tree's overall vitality and ability to reproduce are decreased. Acid rain has contributed to a decline in the annual productivity of coniferous forests in southern Norway, Sweden, and throughout North America. For example, the northeastern United States and eastern Canada saw a significant decrease in the health of sugar maple trees with an accompanying decline in maple syrup production. A survey in 1986 found 82 percent of Canadian sugar maples in decline, with a cost to the maple syrup industry of about Can$88,000,000 in 1985 (Pringle, 1988). (It was later shown that abnormal weather conditions during this period also contributed to the damage [Hall et al., 1998].)

Spruce trees in Germany's Black Forest region were first observed to be dying in the early 1970s. The cause was cadmium and aluminum in the soil produced by the leaching effect of acid rain (Gribbin, 1986). In 1983, West Germany reported that 34 percent of the nation's forests were yellowing and losing needles or leaves. A more thorough survey the following year reported that more than half of the country's forests, about 7 million hectares (18 million acres), had been damaged, including two thirds of the Black Forest. In 1982, Germany committed to reducing sulfur dioxide emissions by 50 percent by 1990. The situation began to improve noticeably.

The earliest sign of injury to trees is in the foliage, while other parts of the plant appear to be more resistant. Eastern white pine is particularly sensitive, showing acute injury in a matter of hours to sulfur dioxide concentrations as low as 25–30 parts per billion (ppb). Many plant species show acute damage within hours to concentrations of 100–1000 ppb, whereas others show acute damage only if sulfur dioxide concentrations are above 1 ppm (Environment Canada, 1990). The effects of acid rain are particularly strong in evergreen forests at high elevations, because acidity is concentrated in fog and at the base of cloud formations (Young, 1990).

A complicating factor with acid rain is that other pollutants also contribute to plant deterioration, placing even greater stress on forests. In the

Appalachians of the eastern United States, for example, research indicates that forest damage is also due to ozone smog (Mitchell, 1991).

Although crop plants have generally only shown minor adverse effects that result from acid rain, research shows that soybeans, an important livestock food and export crop, are seriously affected.

LICHENS AS AIR QUALITY MONITORS

Lichens are composite organisms of fungus and algae, which grow on rocks and trees. Because they have no roots, lichens depend entirely on nutrients absorbed from the air. A study at Brigham Young University demonstrated that lichens function as natural filters, drawing in and accumulating air pollutants. The researchers concluded that lichens could replace expensive environmental monitors, since they accumulate pollutants in the same proportion as they are found in the surrounding air. In another study, dwindling lichen populations in the United Kingdom were found to correlate with rising sulfur dioxide levels in the air, allowing scientists to accurately estimate sulfur dioxide concentrations in the air simply by measuring lichen populations (Natural History Museum, 2004). Lichens are used as an indicator of ecosystem health by the US National Park Service, Forest Service, and Fish and Wildlife Service, as well as other organizations (Blett et al., 2003).

Animals and Humans

When certain elements—aluminum, lead, mercury, cadmium, zinc, and nickel—leach out of soils, they can move into the food chain and become concentrated in the tissues of fish and other animals. In Sweden, moose have such high cadmium concentrations in their livers that the government has warned hunters not to eat their meat. In Canada, raccoons living near acidified lakes have five times the normal mercury concentrations in their livers.

Humans are not exempt from these effects. In the Canadian Shield, acid leaches high levels of lead and copper from rock and soil into drinking water. It also increases levels of methyl mercury in water. This, in turn, concentrates methyl mercury in fish, and then in the people who eat fish. The aboriginal Cree in Quebec have been found to contain abnormally high mercury levels (Pringle, 1988).

Asthmatics and other people with breathing problems appear to be particularly sensitive to sulfur dioxide. For example, tests show that when asthmatics (who were not otherwise displaying any symptoms) were exposed to as little as 0.25 ppm sulfur dioxide for only five minutes during mild to moderate exercise, bronchial constriction, or tightening of the airways, could develop. A study conducted by the Harvard School of Public Health over five years in six cities found exposures to acid air pollutants caused increased respiratory diseases—especially bronchitis—in children. Although young people with a history of asthma and wheezing were the most susceptible, the effects were seen in all children (Mitchell, 1991).

Adults are also affected by sulfur dioxide. Studies have shown that it causes a significant reduction in lung capacity in individuals who exercise (Environment Canada, 1990). Dr. Philip Landrigan of New York's Mount Sinai School of Medicine has argued that "acid rain is probably third after active smoking and passive smoking as a cause of lung disease" (Pringle, 1988).

Structural Damage from Acid Rain

Acid rain also causes damage to paint, buildings, and other structures, especially those built of limestone, which is easily attacked by acid. Irreparable damage has been done to some of the greatest architectural treasures in the world, including the Gothic cathedral in Cologne, the Parthenon in Athens, the Notre Dame cathedral in Paris, and the Statue of Liberty in New York. Acid rain is erasing history and taking a toll on culture. The cost of acid rain damage to buildings is difficult to assess but in the early 1980s the US Army Corp of Engineers estimated it to be about $5 billion per year in the United States.

But there is at least one positive effect of acid in the atmosphere. It has been shown that sulfuric acid in clouds blocks some of the sun's UV radiation. This causes some global cooling, counteracting to a small degree the global warming caused by greenhouse gas emissions.

Solutions to Acid Rain

When the problem of acid rain was discovered, one of the first steps was to make industrial smokestacks taller. Industry spokespersons claimed that the acid-forming compounds would be carried high into the atmosphere and dispersed over such a wide area that they would fall back to the Earth in harmless traces. In 1970, for example, in response to the devastation of

FIGURE 6-1

The Superstack in Sudbury, Ontario (Paul Zizek)

the surrounding area—which had come to resemble blackened barren moonscape—Inco built a "superstack" 400 m (1,312 ft) high at its Sudbury, Ontario, smelter. Unfortunately, this did not fix the problem. Instead, higher stacks only cause acid to be transported further away, from 400 km to over 1,500 km (250–930 mi.); pollutants from the Sudbury superstack produced an increase in acid rain in Europe. It turned out that for acid rain, dilution was not the solution.

The midwestern United States has deep soil deposits and bedrock of limestone and dolomite with good acid-neutralizing capacity; the tall stacks of their power plants, however, send much of their emissions into Canada, where the Precambrian Shield has poor buffering capacity. Ironically, if the midwestern coal-fired plants were to use shorter smoke-stacks, the overall acid rain impact would actually be much less, as much of the acid released would be neutralized nearby.

One way of neutralizing the acidity of lakes is to add lime or limestone. Sweden, for example, uses "wet-slurry dosers" containing fine particles of calcium carbonate (limestone). Unfortunately, this technique is expensive, does not treat soils or trees, and is not feasible for running water. And, in effect, it treats the symptom rather than the problem.

A proper solution to acid rain reduces the emissions of sulfur dioxide and NOx at their source. The main method of reducing acid rain has been to install pollution control devices on the plants that burn fossil fuels. The most common and effective of these are called "scrubbers". (Scrubbers and other methods to reduce sulfur dioxide and NOx emissions are described in the next chapter.)

Another method is to switch from fossil fuels with a high sulfur content to fuels with a lower content such as natural gas or low-sulfur coal. This is not always economically feasible, particularly if the source of low-sulfur coal is far away or if natural gas prices continue to climb.

Because coal is abundant worldwide, and because oil and natural gas supplies are dwindling, active research is underway to find ways of making coal cleaner and environmentally more acceptable (See Chapter 7).

Regulatory Controls

The goal of regulatory agencies was to limit acid deposition to 10–20 kilograms of sulfur per hectare (8.8–17.6 pounds per acre) per year, an amount that scientists feel most ecosystems should be able to absorb. In the early 1980s, deposition in many parts of Europe exceeded 50 kilograms per hectare (44 pounds per acre) (Gribbin, 1986).

In 1990, the US Clean Air Act set a goal of reducing annual sulfur dioxide emissions by nine million tonnes (9.9 million tons) below 1980 levels.

View of Earth including Antarctica seen from Apollo 17 on December 7, 1972

The Earth seen at night from space on December 7, 2002

Clouds and sunglint over the Indian Ocean from the space shuttle Discovery
on September 20, 1973

Cirrus clouds catching the sun at dusk

Smog in Toronto with the CN Tower barely visible

Enormous open-pit coal mine in Germany

Oil field in Iraq

Steel mills in Hamilton, Ontario

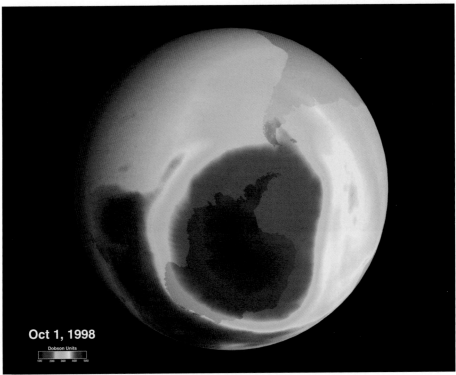

Oct 1, 1998

Dobson Units

100 200 300 400 500

The hole in the Antarctic ozone layer on October 1, 1998

The STS-92 space shuttle astronauts photographed upstate New York at sunset on October 21, 2000. Water bodies (Lake Ontario, Lake Erie, the Finger Lakes, the St. Lawrence and Niagara Rivers) are highlighted by sunglint. The layer of atmospheric pollution is capped by an atmospheric inversion, which is marked by the layer of clouds at the top of the photograph.

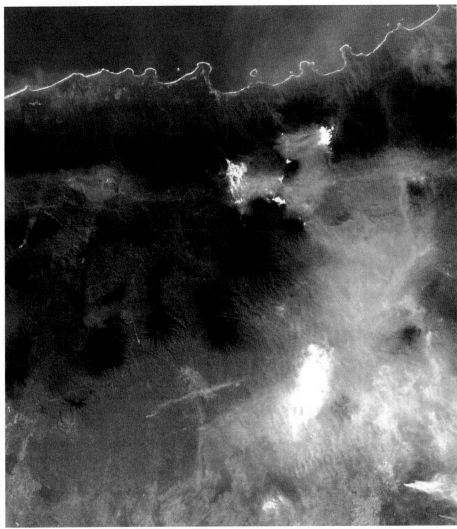

Smoke from forest fires in 1997 over Sumatra, Indonesia, viewed from the space shuttle Atlantis. At one point the smoke covered an area larger than the continental United States.

Forests and streams remain in equilibrium for millennia

Reductions were imposed on fossil-fuel power plants in two phases. Phase I began in 1995, affecting 445 units at 110 mostly coal-burning electric utility plants in 21 eastern and midwestern states. The results were positive and by 1995, emissions at these units were reduced by almost 40 percent below their required level.

Phase II began in 2000 and tightened the annual emission limits for these large, higher-emitting plants. It also set restrictions on smaller, cleaner plants fired by coal, oil, and gas, encompassing over 2,000 units in all. The program applies to existing units with a capacity greater than 25 megawatts (MW) (as well as all new units).

The Clean Air Act also called for a 1.8 million tonne (2 million ton) reduction in NOx emissions by the year 2000. A significant portion of this reduction was achieved by requiring coal-fired utility boilers to install low-NOx burner technologies and meet new emissions standards.

In 1994, the US Acid Rain Program made a dramatic departure from traditional command-and-control regulations by introducing an allowance, or credit, trading system, which harnesses free-market incentive to reduce pollution. Under this system, each utility is allocated a number of allowances, or "caps", based on their historical fuel consumption and emission rates. Each allowance permits a unit to emit 1 ton (0.9 tonne) of sulfur dioxide during or after a specified year. Allowances may be bought, sold, or banked. Anyone may obtain allowances and participate in the trading system. Regardless of the number of allowances a source may acquire, it may not emit at levels that violate federal or state limits.

If a company emits less than its cap, it earns a credit, which it is free to bank for future use, or to sell. If a company emits more than its cap, it must purchase credits to match the amount by which the cap has been exceeded. In this way, industry pays directly for pollution. Furthermore, each company is free to select its own strategy: installing pollution control equipment, switching to cleaner fuels, improving their energy efficiency, or using other methods.

This market-based emissions-trading method provides strong motivation not just to meet the cap, but to get emissions as far below the cap as possible. A company that reduces emissions below the cap generates credits that are worth money. To ensure continuing improvement in air quality, the caps are set progressively lower over time. The acid rain emissions-trading program is similar to the international trading of carbon dioxide emission credits that forms a strategic part of the Kyoto Protocol to reduce global warming (see Chapter 9).

A credit trading system, although simple in concept, requires considerable effort both by the government and industry. For example, determining emission levels for the baseline year requires historical information on plant configuration, air-pollution control equipment, production, and fuel consumption. In addition, estimating current emission levels may require production data, results of stack testing, continuous emission monitoring data, and established emission factors. All these estimates need to be made using industry- and government-accepted methods, and be carefully documented and verified. Verification is an important and challenging task. For major emitters, the quality of these estimates is of critical importance, given the large costs involved.

To ensure the system works properly, the EPA has established an Allowance Tracking System, which records and monitors all allowances that have been issued and all organizations that hold allowances. Between 1994 and 2002, approximately $14 billion was spent on trading credits, representing about 23 million tonnes (25 million tons) of sulfur dioxide emissions (UNEP, 2004). Money is a powerful motivator and the program has proven successful, resulting in decreased costs in pollution control equipment and emission reductions exceeding regulatory requirements.

During Phase II of the program a permanent cap of 8.95 million allowances was established for utilities. This cap firmly restricts emissions and ensures that environmental benefits will be achieved and maintained. The cap may be lowered in future years.

A sure sign of success is that this allowance-trading strategy has been embraced by the private-sector marketplace. In late 2004, the Chicago Climate Exchange, which trades allowances for global warming emissions, expanded its programs to include futures trading on allowances for sulfur dioxide emissions.

Other jurisdictions are embracing this innovative approach. Ontario, for example, introduced emissions-reduction trading for sulfur dioxide and NOx at the end of 2001 for the power generation sector. This was the first emissions trading system in Canada. In 2005, six other industries—steel, pulp and paper, petroleum, refining, cement, and carbon black (a material produced by the incomplete combustion of heavy petroleum products)—were added. The program is slated to further expand not only to other sectors but also to US emitters whose air pollution affects Ontario's air quality.

FUTURE CURRENCY?

Some believe that emission reduction credits will become the currency of the twenty-first century, and will be allocated with the same care as capital funds are today.

In 1985, the governments of Canada and the seven eastern provinces set a target of reducing sulfur dioxide emissions by 50 percent to 2.3 million tonnes (2.5 million tons) by 1994. This program was expanded to include the whole country with a national cap of 3.2 million tonnes (3.5 million tons) per year. This goal was established based on estimates of the natural capacity of the soils, lakes, and plant life in eastern Canada to neutralize the acid being deposited (no more than 20 kilograms of sulfur per hectare [17.6 pounds per acre] per year). This target was met and surpassed, primarily due to the use of lower-sulfur fuel and the installation of scrubbers. By 1994 sulfur dioxide emissions were 54 percent lower than in 1980. By the early 1990s, the area in eastern Canada receiving more than 20 kilograms per hectare (17.6 pounds per acre) per year of acid precipitation had decreased by 61 percent (Environment Canada, 2004b).

Further research in the mid-1990s, however, showed that Canadian ecosystems are more sensitive to acid rain than had been thought. Scientists estimated that an area of some 800,000 km^2 (308,000 sq. mi.)—extending from central Ontario through southern Quebec and across much of Atlantic Canada—was still receiving more acid precipitation than its natural systems could tolerate. Further reductions were needed to protect the environment. A new program was implemented called the "Canada-Wide Acid Rain Strategy for Post-2000". This initiative committed most provinces to reducing sulfur dioxide emissions to 50 percent below 2000 levels by 2010 (Environment Canada, 2004b).

As part of the Acid Rain Strategy, an assessment was conducted in 2004; this assessment concluded that despite substantial reductions in sulfur dioxide and NOx emissions, acidification continues to be a serious problem. Aquatic and terrestrial ecosystems across much of southeastern Canada are still receiving too much acid deposition. Lakes in affected areas generally do not show widespread improvement in acidity (CCME, 2006).

International Efforts

Recognizing that air contamination travels from one country to the other, and motivated principally by acid rain concerns, in 1991 the United States and Canada signed the Canada–US Air Quality Agreement to control cross-border air pollution. The two countries committed to specific targets and schedules in reducing sulfur dioxide and NOx emissions. The United States agreed to reduce its sulfur dioxide emissions by 40 percent from 1980 levels by 2010. Since 1994 the two nations have expanded the agreement to include identification of new sources within 100 km (62 mi.) of the border. The results have been good, and it is expected that the established targets will be met and exceeded.

In December 2000, the agreement was expanded to include the reduction of transboundary flows of NOx emissions from fossil-fuel power stations. In 2000, AIRNOW, the US EPA's real-time air quality program, was extended into Canada in a cooperative program involving the Maritime provinces and New England states. The Commission for Environmental Cooperation of North America (CEC) was established to facilitate cooperation in resolving air pollution problems between Canada, Mexico, and the United States (Environment Canada, 1998b).

In industrialized countries, the threat of acid rain is recognized, and control programs implemented that have substantially reduced the emissions of sulfur dioxide. In 2000, Canada's total sulfur dioxide emissions were 2.5 million tonnes (2.8 million tons)—well below the national cap of 3.2 million tonnes (3.5 million tons) and 45 percent below 1980 emission levels. Further, the trading of sulfur dioxide and NOx allowances is expected to grow and, along with other measures, this innovative concept should reduce these emissions even further (IJC, 2002). In the United States, sulfur dioxide emissions decreased by 35 percent from 1980 to 2002 (IJC, 2002).

On the other hand, emissions of NOx have not improved in either country since 1980. Recent evidence suggests that acid rain damage may be more complex and longer-lasting than was initially believed. Many lakes remain acidified, and many sensitive areas are still receiving acid deposition that exceeds their buffering capacity. Despite the improvements, the problem has not been solved.

The larger concern is that as the global population and economy continue to grow, more energy must be generated. This is clearly demonstrated in the US federal energy plan. In 2001 President George W. Bush stated that America needed 1300 to 1900 new power plants—many of them coal-fired—to be built over the next 20 years. Even with the most modern pollution-control technology, this huge increase in burning fossil fuels will have an enormous and negative impact on our world.

Perhaps the most alarming idea is that most other countries in the world have similar energy goals, and many of them, China included, are more lax in pollution prevention. Is there hope for our endangered atmosphere? Or is this another version of the automobile scenario—global growth outpacing our best technical efforts, and allowing acid rain to quietly tighten its noose on our lakes, forests, and oceans? Are the loons safe?

7

King Coal and Fossil Fuels

Be it known to all within the sound of my voice, whoever shall be
found guilty of burning coal shall suffer the loss of his head.
—King Edward I, during a session of Parliament in the fourteenth century

I have visited a coal mine only once, and it was like a Dickensian nightmare.
Outfitted in baggy blue overalls and with a hard hat jammed on my head,
I boarded a small rail car that rumbled into the depths of a black tunnel. For
two long kilometers (1.2 mi.) we clanged and swayed toward a coal seam
lying far under the Atlantic Ocean. As we passed a rail car transporting a crew
back to surface, the thin beams of our headlamps revealed glimpses of tired
faces caked with black coal dust and with almost alien-looking eyes gleam-
ing as though from inside a deep dank well. It was the early 1980s and I had
come to Cape Breton, Nova Scotia, to witness firsthand the "longwall method"
of coal mining so that my colleagues and I, who were modeling mine behavior using computer software, could better match theory to reality.

As we neared our destination, a noise grew louder. We dismounted
and walked along a tunnel toward the working coal face, and with each
step the noise and reverberation increased until we were enveloped in a
thundering crescendo. We turned a corner and there, with glowering
floodlights throwing long eerie shadows through the dusty darkness, was
the monster: a huge cylindrical machine equipped at one end with ominous rotating blades grinding and shearing chunks from a seam of coal
as effortlessly as if it were cotton candy. The monster was tearing the bow-
els of the Earth asunder, sending a constant torrent of coal pieces flying
and falling onto a huge conveyer belt that carried them away. The shearing machine slowly worked its way back and forth across a face as wide
as a football field, creating a large room as it ripped out the coal.

Over the roar of the mining operations I could also hear—almost feel—
a deep groaning, creaking sound coming from around us. It was a fearful
sound that made me quiver to the core, for it was the sound of rock
formations protesting as they bent and broke. Although I knew the

eventual collapse of the room was planned, the sight—and sound—of geologic formations sagging and buckling made me shiver. We huddled under large steel shields supported by powerful hydraulic legs that held up the roof along the cutting face, providing a safe area for miners to maneuver and service the monster and its conveyor belt. When one row was finished, the hydraulic legs would step forward like a Star Wars robot and lock the roof-support shields into place for the monster to cut the next row.

For weeks afterwards, uninvited images of that nightmarish mining operation would suddenly flash through my mind. I was shocked by the dangerous, dirty working conditions I had witnessed, conditions I thought had disappeared with the long-past Industrial Revolution. But even more mesmerizing was the sheer enormity of the mining operation. The quantity of coal being ripped from that mine was immense. And what I witnessed is just a small fraction of the coal mining taking place around the world! Enormous open pit mines—one in Germany measures 4 by 5 km (2.5 by 3.1 mi.)—rip out prodigious quantities of coal. There are tens of thousands of coal mines around the globe (1,438 just in the United States in 2006 [EIA, 2008b]), all working constantly, dynamiting, grinding, shearing, and tearing this black organic material from the ground.

In 2004, over 15 million tonnes (16.5 million tons) of coal were mined every single day around the world, enough to fill a railcar train approximately 2,860 km (1,780 mi.) long every 24 hours. This daily train with each car loaded with 100 tonnes (110 tons) of coal would stretch from Chicago almost all the way to San Francisco.

Instead of picturing this imaginary train, you can visit the railway track between Gillette, Wyoming, and Alliance, Nebraska, which carries more coal than any stretch of rail in America. You can watch 30 trains pass by every day—more than one per hour—each over 2 km (1.2 mi.) long and carrying more than 9,000 tonnes (10,000 tons) of coal (Goodell, 2006). Coal is indeed a business of huge quantity, and much of the cost of coal is simply in handling and transporting these immense volumes.

What becomes of these massive quantities of coal? Virtually every single lump will be consumed by flames to make steel, generate electricity, or provide some other form of energy.

The "dirty" nature of coal and its generous contribution to global warming are well recognized, yet more coal is burned today than at any time in the past. Many people associate coal with the Industrial Revolution, and think it an energy source whose time has long gone, but that is not the case. Coal is very much in demand, and consumption continues to increase. Between 1999 and 2009, around 250 new coal-fired power plants will be built worldwide; about half will be in China. A further 483 are projected to follow between 2009 and 2019 (Flannery, 2005). Already, China uses more coal than the United States, the European Union, and Japan combined. And it increased coal consumption 14 percent in each of 2004 and 2005. Every seven to ten days, another coal-fired power plant opens somewhere in China (Bradsher and Barbosa, 2006). The size of the coal mining industry and the quantity of coal our society burns simply overwhelms the imagination. No wonder the emissions from coal are quietly overwhelming the atmosphere.

Furthermore, similar gargantuan operations pump vast amounts of oil and natural gas from reservoirs deep below the surface. In 2003, fossil fuels produced about 86 percent of the world's energy, with oil at 38.5 percent being the dominant energy resource. Hydro, nuclear, and renewable energy sources provide the remaining 14 percent. Over the coming decades, the use of oil will diminish while the use of coal will increase.

GROWING ENERGY NEEDS

Humans have an enormous appetite for energy. From 1950 to 2000, the world's population grew 140 percent, but was far outpaced by fossil fuel consumption, which grew 400 percent (Goodell, 2006).

The combustion process transforms fossil fuels, unlocking their energy. But it also releases volatile gases and other impurities that are trapped inside. Coal is especially a problem: roughly one-third of coal consists of volatiles and impurities. Even though modern plants in developed countries have pollution control equipment, large amounts of pollutants enter the atmosphere every single day. Particulates emitted from US coal-fired power plants, for example, are estimated to kill about 30,000 people each year (Freese, 2003).

In addition, the combustion itself—the oxidation of carbon, which unlocks energy—creates carbon dioxide. It happens in such enormous quantities that the world's atmosphere is steadily becoming warmer. The use of coal and the other fossil fuels has reached such a colossal scale that the impacts are no longer local, or even regional, but are having global repercussions.

We are looting nature's resources because we are addicted to the energy that coal and the other fossil fuels provide. Society cannot survive without energy, and we will pay whatever price we must to obtain it. Our homes, which provide comfort and security, are heated by natural gas or fuel oil. Cars, railways, and airplanes, which provide freedom of movement, are powered by gasoline and diesel fuel. Electricity, which drives the thousands of conveniences of modern life, is generated largely by coal, natural gas, and oil. In fact, combustion for electricity generation by utilities is the end use for 92 percent of the coal consumed in the United States (EIA, 2008c). And many kinds of industrial plants, including smelters that process mineral ores and steel mills, use fossil fuels as their basic power source.

Our dependence on fossil fuels grows and grows. And the ever-expanding human population only increases our need. In 2005, for example, a record amount of coal (1,026 million tonnes/ 1,131 million tons) was mined in the United States, and 2006 was another record year with 1,056 million tonnes (1,164 million tons) extracted from the Earth (EIA, 2008b).

All fossil fuels are described in this chapter, but the focus is on coal: this is the fossil fuel that is gaining dominance as oil and natural gas supplies are beginning to dwindle. Also, it is by far the dirtiest fuel, generating more air pollutants than oil or natural gas.

History

Only wood has been used longer than coal for producing heat; archaeological records indicate that coal was burned at least 4,000 years ago. Long before the industrial age, coal cast a somber shadow over the landscape. The soft bituminous coal introduced in England in the 1300s as a substitute for wood to heat homes and shops created such annoying quantities of soot and noxious sulfur dioxide that King Edward I issued an edict prohibiting London merchants from burning coal (Officer and Page, 1993). One person caught violating this edict was put to death.

It was not until the steam engines and factories of the Industrial Revolution in the eighteenth and nineteenth centuries, however, that coal became the world's major energy source. But times have changed and modern society no longer uses coal to heat homes (although this practice continues in many developing countries), drive trains, or power looms or other machinery. Coal's importance continues today, however, with the primary purpose of generating electricity.

On September 4, 1882, Thomas Edison threw a switch in lower Manhattan in New York that was to transform society. He had designed and constructed the world's first coal-fired power plant, called "Long-Legged Mary Ann". When the electrical circuits closed, lights went on in the buildings in several surrounding blocks. This was a technological triumph of the first order for it allowed energy to be unlocked from a common and abundant rock (Goodell, 2006).

From that small beginning, the use of coal to generate electricity has grown enormously. The growth has not always been steady, for nuclear power and natural gas have on occasion challenged King Coal's supremacy. But coal has outlasted its competitors, for it has one overwhelming advantage: it is the most abundant fossil fuel on Earth. Reserves of coal are expected to last for a few more centuries, whereas worldwide production of oil and natural gas are already nearing their peaks.

Most Americans have no idea how much they depend on coal. With the threat of global warming, one might expect coal's dominance to be receding, but it is not. Coal continues as an almost unimaginably huge pillar of the power and industrial sectors. In the United States between 1970 and 2000 the amount of coal-generated electricity tripled, and it continues to grow robustly. The United States' per capita consumption of coal is three times that of China. It is more central to the economy than it has ever been before, yet there is little connection in most people's minds between the light switch on the wall and the dirty black rock (Goodell, 2006).

There is good reason for America's dependence on coal; after all, it is the Saudi Arabia of coal. Saudi Arabia holds 20 percent of world's oil reserves, while the United States has more than 25 percent of world's recoverable coal reserves, with estimates at about 245 billion tonnes (270 billion tons). In comparison, China has about 114 billion tonnes (126 billion tons). As coal industry executives proudly announce, enough coal lies under American soil to last 250 years at present consumption rates (Goodell, 2006).

This bounty of coal has many implications. One of the most significant of these is that the United States will not "freeze in the dark" after oil becomes scarce. The United States can always turn to coal, no matter what the cost. And the cost is significant: coal-fired power plants cast into the atmosphere nearly 40 percent of US emissions of carbon dioxide, not to mention a plethora of other pollutants. In fact, America's transition to coal has already begun. The production of oil in the United States peaked in about 1970, and in 2006 it imported two-thirds of the oil it consumed. Coal has quietly made a resurgence in the United States, especially since about 2000.

Not surprisingly, the coal industry, in its quest to increase profits, views the Kyoto Protocol as an international plot by jealous competitors to thwart one of America's great economic advantages and to cripple America's "energy independence". In a land where lobbyists have inordinate input into government policies, the US coal lobby is especially powerful.

COAL AS KING MAKER

In the 2000 presidential election, Al Gore, the Democratic candidate, was taking global warming seriously and promised to reduce carbon dioxide emissions. This was heresy to the coal industry, which—envisaging the unpleasant prospect of curtailed growth—threw its enormous clout behind George W. Bush. Intense lobbying helped deliver key industrial states to Bush, including West Virginia—a coal-mining state that had not voted for a Republican presidential candidate in 75 years. Although hanging chads and other voting irregularities in Florida are usually the reason given for President Bush's razor-thin victory, it can just as reasonably be attributed to the five electoral votes delivered by West Virginia, thanks in large part to the coal industry (Goodell, 2006).

President Bush did not forget his new friends. Once he was in office, regulatory agencies were staffed by former coal industry executives and lobbyists, and Bush's energy policy called for hundreds of new coal-fired plants. And the Kyoto Protocol was never signed by the United States, although polls showed 60 percent of Americans were in favor.

Because of its overwhelming abundance, coal has enormous potential to supply energy far into the future. But it will only be acceptable if its negative features can be overcome.

Formation

Let us look at how fossil fuels are formed. Oil, also called petroleum, consists of naturally occurring liquid hydrocarbons, which when distilled and cleaned of impurities yield a range of combustible fuels, petrochemicals, and lubricants.

Oil was formed when ocean-dwelling, single-celled plants called "phytoplankton" were buried in sediments at the ocean floor. Oil formation requires the bottom water to be stagnant or poorly ventilated so that oxidation does not occur. Most of the world's oil deposits are thought to have formed in deep, still, oxygen-poor basins after upwelling in the oceans brought nutrient-rich water from the depths to the warm surface, causing phytoplankton to bloom in enormous quantity. When they died, they fell to the bottom and accumulated without being consumed by bacteria. There the phytoplankton masses were trapped in predominantly clay sediments, where they began to transform into hydrocarbons. Ideal conditions for hydrocarbon formation are temperatures between 20°C and 90°C (68°F–194°F) at depths between 600 and 6,000 m (1,969–19,685 ft). If the temperature is too high, only natural gas is created. Once the hydrocarbons have formed, they can migrate both upwards and sideways until they enter a porous reservoir rock from which further migration is prevented by surrounding impermeable layers (Emiliani, 1991). Oil and gas have a low density, so if the reservoir contains groundwater, they will float above it.

"Natural gas" refers to hydrocarbons that exist as a gas under ordinary conditions. Methane is the main constituent; however other compounds may be present, including ethane and propane, along with various other possible impurities. Natural gas may occur alone or be associated with oil.

Coal was formed millions of years ago by the burial and compression of trees and plants. Much of the coal in the world formed during a time of cyclic glaciation that occurred during the Pennsylvanian period (also known as the Carboniferous period, due to the massive coal formation) from about 360 to 290 million years ago. At this time, all the continents were joined together in one huge supercontinent called Pangea. Each time ice built up in the southern hemisphere, the sea level dropped and river deltas in the northern hemisphere built up flood plains with rich soil that nurtured forest growth. When the southern ice melted, the northern lowlands were flooded, killing the forests. The forests were buried and compressed under thick layers of marine sediments, and subsequently formed coal. In West Virginia, more than 100 coal seams, each one representing one such cycle, have been measured. This coal-forming process, known as the "first coal age", took place over a period of more than 40 million years and yielded mostly anthracite and bituminous coals (Sherman, 2004; Emiliani, 1991).

This coal-making era was an important milestone in the planet's evolution, for it marked the completion of the atmosphere as we know it today. The vast forests of trees and plants that now lay underground as coal deposits had devoured carbon dioxide from the air, replacing it with oxygen, so that the oxygen concentration reached its present level of about 21 percent. That we should be burning coal (and other fossil fuels) in such enormous quantities that we are slowly returning the atmosphere to its primordial composition of 300 million years ago is a monumental feat.

The formation of coal deposits continued but on a quieter scale in the "second coal age", which includes the Cretaceous (144 million to 66 million years ago) and Tertiary (66 million to 2 million years ago) periods.

TIME SCALES

It is almost impossible to imagine the slow ponderous movements of nature during the "coal ages". A tiny seed took hold in fertile soil. A small sapling grew into a mature tree. It flourished for decades, perhaps for a few centuries, constantly breathing in carbon dioxide and exhaling oxygen. Eventually it toppled and decayed, contributing to the complex life of the forest by forming food and a home for insects, animals, moss, and other plants. Slowly it decomposed into soil and peat. As the land subsided over millennia, its remains—along with those of millions and millions of other plants and trees—were slowly covered with sedimentary ooze and were compacted below an ever-deepening mass. Over countless millennia this organic mass transformed until only a seam of coal remained where once had stood a proud forest. Then the cycle began again, and again, and again. In this patient fashion, over millions of years, nature formed the black rock we call coal.

Now we humans are ripping these coal seams out of the Earth and destroying them in the furnaces of industry in what, on the timescale of the Earth, is a mere fraction of a heartbeat. It's as if someone had lived frugally, carefully saving money all their long life, and then lost it all in five minutes at the casino.

In 2007, the proven coal reserves of the world were estimated at 847 billion tonnes (934 billion tons) (British Petroleum, 2008). Their distribution by continent is: Europe and Eurasia, 32 percent; North America,

30 percent; Asia and Pacific, 30 percent; Middle East and Africa, 6 percent; and South and Central America, 2 percent.

Coal in North America is found along the Appalachian mountain chain, particularly in western Pennsylvania, West Virginia, and Kentucky—probably the single largest coal reserve in the world—and also in Nova Scotia. Coal (and oil) is also found in the Michigan Basin, to the north of it, and in the Indiana and Illinois basins to the south. The western coalfields lie in sedimentary basins to the east of the Rocky Mountains. Bituminous coal occurs in the Raton Basin and the huge Williston Basin, which extends farther north, but contains rather low-grade coal. Farther north lies the vast Alberta Basin, with coal exposed in the foothills of the Rockies; this basin also contains one of the world's largest coal deposits (*Britannica*, 2003). As a coal seam is squeezed and heated by surrounding geologic formations (geologists call this "metamorphism"), volatiles are progressively squeezed out and the coal becomes denser and purer, or higher in "rank". Coal is classified by type (differences in plant materials), rank (degree of metamorphism), and grade (amount of impurities). Table 7-1 shows the different ranks of coal and some of their properties; it is seen that density and heat of combustion increase with rank.

Coal is readily combustible and contains more than 50 percent by weight and 70 percent by volume of carbonaceous material. But here lies the problem: coal is a complex substance and, in addition to carbon, contains a large range of organic and inorganic compounds such as 20–35 percent by weight of volatiles; sulfur; metals such as mercury, nickel, and arsenic; and naturally radioactive compounds. These are all released when

TABLE 7-1: Rank of Coal (after Emiliani, 1991)

Name	Carbon (% dry)	Heat of Combustion (calories/gram)	Bulk Density (grams/cm^3)
Peat	40	3,000	–
Lignite	50	4,000	0.70
Sub-bituminous	60	6,000	0.75
Bituminous	80	8,000	0.80
Subanthracite	90	8,500	0.85
Anthracite	95	7,900	0.90

coal is burned, and in the absence of proper pollution control equipment will enter the atmosphere. Even modern technologies do not capture 100 percent of these emissions.

The volatile substances include a range of compounds: tars; light aromatic compounds like benzene, toluene, and xylene; phenol; cresols and similar compounds; methane; ethane and carbon monoxide; ammonia; hydrogen sulfide; cyanide; and water vapor. The content of volatile matter decreases with increasing rank. In general, coals with high volatile-matter content ignite easily and are highly reactive in combustion applications.

In some industrial processes (such as steel-making) a fuel with high carbon content is required. This is achieved by "coking", or "high-temperature carbonization", in which coal is heated to temperatures between 900°C and 1,200°C (1,650°F–2,200°F). Practically all the volatile matter is driven off as gases or liquids, leaving behind a residue that consists principally of carbon with minor amounts of hydrogen, nitrogen, sulfur, and oxygen.

Coal Mining

The mining of coal from surface and underground deposits today is highly productive and mechanized. Open-pit mines in particular are super-sized industrial operations that are far more invasive than drilling for oil and gas. The largest mines in the United States are open-pit mines in the Powder River Basin of Wyoming. For example, the Cordero Rojo mine owned by Kennecott Energy (a subsidiary of Rio Tinto) sprawls for over 2,600 hectares (6,420 acres) outside the town of Gillette (Goodell, 2006). Entering it is like entering a land of giants. The machinery is overwhelming, with trucks whose tires dwarf a human.

But the massive size of coal mining also brings massive environmental and health impacts. Because surface mining makes large scars on the landscape, reclamation is an important part of the overall mine operation. Reclamation includes backfilling excavations, regrading the surface, and restoring and revegetating the land to reduce erosion, sediment discharge, slope instability, and water-quality problems.

Underground coal mining has always been one of the most hazardous occupations in the world and continues to pose many safety and environmental problems. This is due in large part to the explosive nature of methane gas and coal dust and, to a lesser extent, the instability of the mined openings. For example, there were 361 deaths at Monongah, West

Virginia, on December 6, 1907. And in recent times the risks continue, albeit in developed nations at lower levels. In 2001, for example, 13 men were killed in a methane explosion in an Alabama mine (Goodell, 2006). About 300 die in Ukrainian coal mines each year. On June 21, 2002, an explosion ripped through a Chinese coal mine, killing 111 people. But this was only the tip of the iceberg. China accounts for about one-third of global coal production and due to its often-primitive mining methods accidents are common, with over 6,000 coal mine workers dying each year (Goodell, 2006). Canada has suffered its share of devastating coal mine explosions, the most recent claiming 26 lives on May 9, 1992, at the Westray Mine in Nova Scotia.

In 1983, there were 261 uncontrolled fires in underground coal mines in the United States; once begun, these fires can take literally decades to extinguish.

There are also many deaths and illnesses from occupational exposure to coal dust. The respiratory disease pneumoconiosis or "black lung" is devastating. It is caused by breathing excessive amounts of coal dust, which becomes imbedded in the lungs, causing them to harden through progressive massive fibrosis. Although black lung has been largely eradicated through the use of proper mine ventilation, over 1,500 miners (mostly retired) continue to die from it every year (Goodell, 2006).

In the Appalachian Mountains paralleling the eastern seaboard of the United States, the coal industry has introduced a new way of mining. "Mountaintop removal" offers a way to be ever more productive and to compete with easier-to-mine coal from the western states. Instead of removing the coal from the mountain by sinking expensive mines, the mountain is removed from the coal. This involves moving prodigious amounts of soil and rocks by effectively lopping the tops of mountains and filling in valleys. Of course, coal extracted in this way is not cheap. In this case, the price includes the destruction of more than 1,100 km (680 mi.) of streams, the pollution of rivers and groundwater, and the transformation of about 160,000 hectares (400,000 acres) of rich temperate forest into flat, barren wasteland. A federal report projected that over the decade starting in 2005 about 5,700 km² (2,200 sq. mi.) of land would be affected by mountaintop mining (Goodell, 2006).

Mountaintop mining has been greatly aided by President Bush's connections with the coal industry. In 2004, for example, the debris removed by mountaintop mining was reclassified from "waste" to "fill"—even though this debris is known to leach acid and heavy metals into nearby waterways. Previously, mining was not allowed within 30 m (98.4 ft) of a stream; now the "fill" from mining covers valley-bottom streams. This reclassification has had a huge impact by undercutting legal challenges to mountaintop mining and paving the way for new mines (Warrick, 2004).

Fossil-Fuel Power Generation

Electricity is generated by all three fossil fuels. The fuel is burned in a boiler that is lined with tubes containing water. The heat causes water to boil, just like a kettle on a stove, forming steam; this drives a turbine that turns a generator to create electricity. The steam from the turbine is condensed back to liquid by cooling, usually using water from a nearby river or lake, but sometimes using air in large towers, before it is returned to the boiler. As in all fossil-fuel power plants, only about 35 percent of the potential energy of coal is converted into electricity. The remainder is released into the environment as heat. Various methods of increasing plant efficiency are being actively explored.

Coal is the most frequently used fossil fuel, accounting for 40 percent of worldwide electricity generation. In 2005, about 600 coal-fired power plants produced about 50 percent of the electricity in the United States (National Mining Association, 2007).

In Canada, there are 24 coal-fired power plants, which account for 17 percent of electricity production in the country. Alberta, Saskatchewan, and Nova Scotia are the most dependent on coal. Ontario has the two largest coal plants in the country: Nanticoke on the north shore of Lake Erie and Lambton near Sarnia (1,975 MW).

The Nanticoke Generating Station is Canada's largest coal-fired power plant with its eight units producing 3,920 MW of electricity. I tried to visit Nanticoke Generating Station to see firsthand a coal-fired power plant and its pollution control equipment. But the owner, Ontario Power Generation, denied me access. Perhaps because many including the provincial Liberal Party (2003) consider Nanticoke the dirtiest power station in North America, generating the pollution equivalent of 3.5 million automobiles every year.

Steel Making and Other Industrial Users of Coal

Steel is critical to the economy. It is the building block for an enormous variety of goods including cars, steel girders, cookware, and filing cabinets. In the basic steel-making process, enormous amounts of coal are first converted to coke in coking ovens. The coke is fired in a blast furnace together with iron ore and limestone to create "pig iron". This material is then refined in a separate furnace to form various grades of steel. The coking ovens release volatiles from the coal, creating considerable emissions in the atmosphere.

Technological improvements have allowed coke consumption in blast-furnace iron-making to be reduced to one-third of what it was 80 years ago. Current research is developing direct-combustion technologies that will reduce iron ore directly to iron in a single process. The next generation of iron-making furnaces is likely to operate without coke.

Cement manufacturing is another industry that consumes enormous amounts of energy, with coal often the preferred fuel due to its abundance and cheap cost. High-temperature kilns "bake" limestone and shale to produce cement.

Air Pollution

Pollution from coal knows no borders, and causes international acrimony. Ontario, for example, complains that as much as 50 percent of its smog comes from US coal-fired power stations in the midwestern United States and the Ohio Valley. In 2003, New York State filed a petition with the Commission for Environmental Cooperation, which was established as part of the North America Free Trade Act, for Ontario to clean up the emissions from its coal-fired power stations. The petition claimed that Ontario's aging power plants are among the leading contributors to acid precipitation in New York and the northeastern United States, with the pollution having a major impact on air quality and the health of New York residents (White, 2003).

COAL IS DEADLY

The Harvard School of Public Health conducted a study on the health effects of nine coal-fired power plants in Illinois (HSPH, 2001). It concluded that the emissions cause an annual risk of 300 premature deaths, 14,000 asthma attacks, and more than 400,000 daily incidents of upper respiratory symptoms among the 33 million people living within 400 km (250 mi.) of the plants. The study indicated that if the plants were equipped with modern pollution control equipment, health impacts would be significantly reduced (100 premature deaths, 4,000 asthma attacks, and more than 100,000 daily incidents of upper respiratory symptoms).

The World Health Organization estimates that in India and South Korea about 355,000 people die each year from urban outdoor air pollution, caused mainly by burning coal (Goodell, 2006). As we saw in Chapter 6, acid rain from burning coal is also a serious problem in China, where limestone buildings are dissolving and some ancient statuary has been enclosed in sheets of acrylic plastic to protect it from the acidic air.

COAL TIME-OUT

The Sarnia area of Ontario has a large number of chemical industries in addition to the Lambton coal-fired power station. Occasionally the emissions combine with weather conditions and contaminants blowing in from the United States to cause unacceptably high air pollution levels. The Ontario Ministry of Environment, together with a local industrial society, has established a meteorological warning system. When alerts are posted, major industries must curtail their production or switch to lower-sulfur coal until the alert is removed (MOE, 2001).

Coal-fired power plants are the largest emitters of mercury in the United States, spewing out about 44 tonnes (48 tons) per year into the atmosphere. Exposure to humans is mostly through eating fish, which accumulate mercury in their tissues (Goodell, 2006).

In 2005, the US Environmental Protection Agency made a ruling that demonstrated President Bush's dubious connection with the coal industry. It removed coal-fired power plants from the list of sources of hazardous air pollutants—in spite of the known health hazards of the many toxins they emit that include mercury, particulate matter, volatile organic compounds (VOCs), and heavy metals. Instead of facing tough restrictions on emissions, the coal-fired plants now have to comply with softer, more flexible standards (Goodell, 2006).

Towards Cleaner Coal

Coal poses a threat to the ecosystem for two reasons: it contains many pollutants, and it is burned in enormous quantity. Consider sulfur, for example. In 2004, about 5,500 million tonnes (6,060 million tons) of coal were burned worldwide, releasing about 110 million tonnes (120 million tons) of sulfur (assuming an average sulfur content of 2 percent). Even if scrubbers captured 90 percent of this (an overestimate, as China and India have few state-of-the-art pollution controls), about 11 million tonnes (12 million tons) of sulfur escaped into the atmosphere. This represents an enormous atmospheric load and it increases year upon year. Acid rain will continue to be a growing problem.

A relatively simple method of decreasing emissions from a coal-fired plant is to use a lower-sulfur coal, oil, or natural gas. Natural gas, in particular, has significant environmental advantages. It burns more cleanly, emitting much lower quantities of nitrogen oxides (NOx) and carbon dioxide, and virtually no sulfur dioxide or other hazardous compounds such as mercury. In addition, natural gas turbines can generate electricity with an efficiency of 40 percent (compared to 34 percent for coal stations), helping to offset the higher cost of gas.

Another way of decreasing fossil fuel consumption is by making power plants more efficient. The "combined-cycle gas turbine" is a new technology with a significantly higher efficiency of 58 percent. It uses exhaust from the gas turbine—hot enough to produce steam—to drive a steam turbine and generator. Today, with natural gas supplies dwindling, combined-cycle gas turbines are the preferred technology for new electricity plants using natural gas, even though they require a large capital investment (about $750 per kilowatt).

Fossil fuel stations can be made more efficient in another way: approximately 66 percent of the heat left over after generating electricity can be used for heating nearby buildings or powering nearby industry. This is called "cogeneration". Although common in Europe, unfortunately cogeneration is practiced by virtually no coal-fired power stations in North America.

The dark side of coal is well recognized and research has long been seeking ways to minimize emissions and environmental impact. Efforts are directed at three areas: the processing of coal after it has been mined, adding pollution control technologies to coal-fired stations, and developing combustion systems that burn more cleanly.

Coal Processing

Researchers are looking for ways to make coal cleaner even before it reaches a combustion chamber. Sulfur in coal is either chemically bound to the carbon (organic sulfur) or is combined with iron particles as pyrite (FeS_2) that are distinct from the coal itself (pyritic sulfur). In the latter case, the coal can be treated to remove the pyrite. In "coal washing", for example, coal is crushed, filtered, and washed using froth-flotation techniques that rely on the fact that pyrite is much denser than coal. By grinding the coal into minute particles measuring only a few microns across and then using centrifugal separation, more than half of the pyrite can be removed. The cleaned coal can be reformulated into pellets for easy handling (Corcoran, 1991).

Research is also being conducted into more esoteric methods of coal cleaning. In one method, for example, coal is immersed in heated sodium hydroxide for a few hours and then washed and filtered. Researchers are hoping to some day use biological agents that will "eat" sulfur. Discovering the right "bugs" and learning how to control them is a challenge (Corcoran, 1991).

Pollution Control

In the early days, coal-fired power plants operated with virtually no pollution controls whatsoever. The first step forward was the introduction of scrubbers to remove sulfur dioxide from exhaust gases before they disappeared "up the stack". Scrubbers have become a very effective weapon in combating acid rain. Initially, the British coal industry claimed it would be impossibly expensive to install scrubbers on a large scale. Jeff Gooden's book *Big Coal: The Dirty Secret Behind America's Energy Future* describes how the American coal industry has also fought vigorously against adding expensive pollution control equipment or making costly technological innovations. Nevertheless, progress has been made. In 1987, only about 15 percent of US coal-fired power plants used them; today, they are standard equipment.

Scrubbers (also called "desulfurization units") can be wet or dry (actually, semi-dry). The wet type is used about 90 percent of the time, and involves injecting a lime solution or slurry into exhaust gas. Reaction of the sulfur dioxide with the calcium produces calcium sulfate, also known as gypsum, and removes about 90 percent of the sulfur dioxide. A benefit of the wet process is that fine particulate matter and trace metals, including mercury, are also removed. If the gypsum is pure enough it may be used to make wallboard or drywall. A disadvantage is the relatively large amount of energy needed.

The dry system involves the injection of a fine powdered lime (CaO) or another source of calcium into the exhaust gas. Reaction products are collected together with fly ash using an "electrostatic precipitator". The dry systems are less expensive than wet ones, but are less efficient at removing sulfur dioxide. Also, the reaction products have no commercial value and must be disposed of in landfills (MOE, 2001).

Some scrubbers can fit inside existing ductwork, which is preferable for older and smaller plants with limited space. The drawback is that sulfur dioxide removal efficiency is lower, at about 50–70 percent. The larger, separate scrubbers can remove up to 90 percent of sulfur dioxide.

Particulate matter, including the reaction products from scrubbers, is captured by either a "baghouse" (a series of cloth bags similar to large vacuum cleaner bags through which gases—but not particulates—can pass), or by an electrostatic precipitator in which the flue gas particles are ionized and then captured on large plates with an applied voltage. Baghouses clog easily, so they cannot deal with wet material. To operate efficiently they must be placed before a wet scrubber, followed by a dry scrubber. Since they collect more particulate matter and are better at trapping very fine particles, baghouses are becoming the preferred pollution control technology.

NOx, one of the precursors to smog, form in two ways in coal-fired stations. First, nitrogen that occurs naturally in the air reacts with excess oxygen at high temperatures in the boiler; known as "thermal NOx", this represents about 25 percent of the total. Second, nitrogen chemically bound in the coal reacts with excess oxygen in the boiler; this "fuel NOx" represents about 75 percent of the total.

Low-NOx burners work using a "during-combustion" process in which the amount of air allowed into the primary combustion zone is decreased, creating a fuel-rich, oxygen-poor environment. The absence of oxygen makes it difficult for nitrogen to form NOx. The flame temperature is also kept lower, further suppressing NOx formation. To complete the burning of coal, more air is added after the primary combustion zone—here the temperature is sufficiently low that additional NOx formation is minimized. Another way of achieving lower oxygen ratios is to include 5–15 percent natural gas. This has been shown to reduce NOx formation by as much as 25 percent.

Further emission reduction can be achieved by a technique called "overfire". This system uses even less oxygen in the primary chamber and then injects more air at higher elevations in the boiler to complete combustion, again at lower temperatures. Low-NOx burners are installed on about 75 percent of the United States' coal-burning power stations and typically achieve 30–65 percent NOx reduction. They are the primary technology currently used in the United States to meet NOx regulations (MOE, 2001).

"Selective non-catalytic reduction" is a post-combustion process. Here, ammonia (NH_3) or urea (CON_2H_4) is injected into the exhaust gases and react with NOx to produce nitrogen and water. The technology removes about 60 percent of NOx but releases a small amount of ammonia into the air. "Selective catalytic reduction" uses much the same process. It injects ammonia or urea into the exhaust gases in the presence of a ceramic or metal catalyst. This method works like the catalytic converter in a car and achieves NOx reductions of up to 90 percent.

At this time, there are no commercially available technologies for removing carbon dioxide from coal-fired exhaust gases. This topic is discussed further in Chapter 9. The only practical method currently available for reducing, but not eliminating carbon dioxide emissions is to switch from coal to natural gas.

Advanced Combustion Systems

Coal emissions can also be reduced by replacing standard boilers with more advanced combustion systems. "Fluidized-bed combustion", a process which is more widely used in Europe than in North America, has much potential. Pulverized coal is burned in a bed of powdered limestone with jets of air coming up from below to create a "fluidized bed" in which mixing creates more complete burning than in standard furnaces (DOE, 2001). The sulfur also reacts efficiently with the limestone for high sulfur removal. Because this furnace operates at low temperatures, less NOx are produced. Pollutants are removed inside the combustion chamber so no scrubber or post-combustion controls are needed. Fluidized-bed systems reduce sulfur dioxide by 90–95 percent and NOx by 90 percent or more.

Another combustion technology is the "limestone injection multistage burner", in which limestone is injected into the furnace where crushed coal burns. Nearly all the sulfur dioxide is converted to gypsum, which can be sold. Most of the NOx are removed in a second combustion zone. This system requires about 25 percent of the space required for a scrubber, is less costly, and can more easily be added to existing stations than scrubbers.

"Integrated gasification combined-cycle" (IGCC) plants promise to be a major technological breakthrough and may create the cleanest and most efficient coal-fired power plants (DOE, 2001). The system uses two separate processes. In the first step, called "gasification", crushed coal is mixed with water and oxygen. Then it is heated in a high-temperature, high-pressure vessel to produce methane gas. The methane is burned to drive a turbine that creates electricity. A solvent removes 95–99 percent of the sulfur dioxide from the methane before it enters the turbine. What makes this system so attractive is that the exhaust gas from the gas turbine is hot enough to boil water and generate steam. The steam drives a second turbine and creates even more electricity. For this reason, the system is named "combined cycle". A pilot plant constructed in Daggett, California, in 1984 has achieved efficiency levels of about 55 percent—20 percent higher than conventional coal-fired power stations. With further technological developments it is expected that the combined-cycle systems could achieve an unprecedented 70 percent efficiency. These systems remove about 95–99 percent of the sulfur dioxide and NOx emissions, and the increased efficiency reduces emissions of the greenhouse gas carbon dioxide. Another advantage is that removing carbon dioxide is far easier here than it is at a conventional coal-fired plant, allowing it to be sequestered in deep aquifers (underground formations that contain water).

The US coal-power industry, which is making good profits with the old, proven plants, has been slow to adopt this new technology; in 2007 there were only two small coal IGCC plants operating. The Polk IGCC facility, producing 260 MW of electricity and owned by Tampa Electric, was commissioned in 1996 near Tampa, Florida, and is considered among the nation's cleanest, most efficient, and most economical power-generation units. The Wabash River Coal Gasification Repowering Project outside West Terre Haute, Indiana, started full operations in 1995 and generates 292 MW of electricity.

The Future of Coal

The ideal solution for the planet and its growing population of energy-hungry humans would be to have an unlimited supply of a completely clean energy source. Coal, because of the enormous reserves located throughout the world, satisfies the first requirement. For example, in 2003 Canada had an approximately 800-year reserve of coal, compared to an 11-year confirmed supply of natural gas (Wright, 2003).

The difficulty is the second requirement: a pollution-free energy source. Removing carbon dioxide from the emissions is a particularly vexing problem. Considerable research is being directed at this issue, and recent studies suggest this goal may be realizable. However, commercial production is still a few decades in the future, and then it will take a few more decades to replace existing coal plants with the new installations.

One proposed zero-emission technology is being pursued in the United States (Ziock et al., 2001). In the first stage, a coal and water slurry along with lime (CaO) is gasified and converted into hydrogen gas and calcium carbonate ($CaCO_3$). The process is written like this:

$$CaO \text{ (lime)} + C \text{ (coal)} + 2H_2O \text{ (water)} \longrightarrow CaCO_3 + 2H_2$$

The hydrogen is then used in high-temperature solid-oxide fuel cells to produce electricity, and a fraction of the hydrogen is returned to the first stage. Carbon dioxide formed in the first stage combines with the lime to form calcium carbonate ($CaCO_3$). Waste heat from the solar cells is used to heat (or "calcine") the calcium carbonate. This breaks down to produce lime (which is used again in the first stage) and an almost pure stream of carbon dioxide.

$$CaCO_3 + \text{heat} \longrightarrow CaO + CO_2$$

It is proposed that the carbon dioxide be reacted with magnesium silicate rocks such as serpentine or olivine to produce magnesium carbonate, a solid, inert natural material that can be disposed of in mines or quarries. Studies indicate this would be a technically feasible approach for sequestering the carbon dioxide (O'Connor et al., 2000); it is certainly better than burying it in gaseous form. Other pollutants from the plant—such as nitrogen, sulfur, heavy metals, and particulates—would be in relatively easy-to-manage liquid or solid forms.

Calculations show that the efficiency of converting coal energy to electrical energy by this method should be about 70 percent. That is almost double current rates.

In 2003 President Bush proposed a bold venture called FutureGen, a plant that produces hydrogen and emits no carbon dioxide or pollutants. Even if it is completed on schedule by 2013—an unlikely prospect— several more decades will pass before the technology is streamlined and such plants can be constructed routinely. By then, the battle with global warming may already be lost (Goodell, 2006).

In Canada, the Canadian Clean Power Coalition, an association of coal producers and coal-fired electricity generators, in collaboration with the California-based Electric Power Research Institute, plans to build a full-scale, coal-fired demonstration plant to be in operation by 2012. The goal is to develop a coal-fired plant as clean as a modern natural gas-fired gas turbine plant. At the same time, it would maintain a level of efficiency at or above current levels, be competitive (in terms of cost) with other generation technologies, and have the capacity to capture carbon dioxide emissions.

Every minute of every day machines are ripping coal from the depths of the Earth. Trains carry vast amounts of this black rock to tens of thousands of furnaces. Smokestacks send fumes into the atmosphere, whose defenses are overwhelmed. Although some measures to protect the environment are being taken—especially by developed nations—the scale of coal production is enormous and continues to grow even larger. It is my fear that the outcome will mirror the situation with the automobile and that any technological improvements that are implemented will simply be swamped by ever-increasing growth.

In the next chapter, we look at the story of chlorofluorocarbons, or CFCs. Although CFCs, a synthetic gas, are very different from coal, the two stories have striking similarities. Such copious quantities of CFCs were used that the stratosphere was overwhelmed, causing potentially catastrophic damage. The need for such immense quantities of coal and CFCs is driven, of course, by the demand created by only one species out of the millions that inhabit this globe: *Homo sapiens*.

8

The Tenuous Ozone Shield

I have fond memories of the golden beaches of Australia when I was studying for my doctorate in the 1970s. One perfect turquoise wave after another rolled in from the vast reaches of the Pacific Ocean and crashed and swirled on the shore. Crowds of people basked under a cloudless blue sky at Bondi, Manley, and Coogee beaches where miles of golden sand form sensuous curves between rocky headlands. We played for hours in the tumbling waves, some "hanging ten" on boards and others body surfing. Our only protection from the hot solar rays was bathing suits and a few dabs of zinc cream on our faces.

Now all that has changed. Today, many beachgoers wear hats and long-sleeved shirts. And those who are baring their skin—generally the younger set—are lathered in sunscreen, their skin glistening and oily under the baking sun. Even in the hedonistic lives of Sydneysiders there is an awareness that the environment has changed and the sun is now to be feared as much as worshiped.

There is good cause for concern—and not only among beach lovers—for the ozone layer high in the stratosphere, which offers protection against the sun's UV rays, has grown thinner and developed a dramatic hole above the South Pole. Life on Earth would not be possible without the ozone layer, and now that shield has weakened.

Truly shocking is that this devastating damage to the ozone layer has been caused, not by some enormous cataclysmic event, but instead by everyday, seemingly harmless actions. Who would have thought simply applying hairspray, switching on an air conditioner, or spraying a mosquito could cause global repercussions? But these tiny actions repeated millions and billions of times have had an enormous—and totally unpredicted—impact. The results could have been an ecological catastrophe.

Although they came as a complete surprise when first discovered, the holes in the ozone layer and their underlying causes are now well understood. Furthermore, measures have been taken that should fix the problem. It is a fascinating story complete with drama, conflict, and suspense.

As recently as the early 1970s—when I was basking in the Australian sun—the threat to the ozone layer was entirely unknown. By the late 1970s, the problem had emerged; but everything was theory and speculation, and those who were responsible for manufacturing the dangerous ozone-depleting chemicals fought hard against any changes. But let us start at the beginning and look at ozone chemistry and the ozone layer.

Ozone Basics

Ozone is a blue gas with a distinctive pungent odor consisting of three oxygen atoms (O_3). Unlike oxygen gas, which has only two oxygen atoms (O_2) and is stable, ozone is highly reactive and tends to combine with other molecules and atoms.

In 1881, British teacher Walter N. Hartley discovered that ozone absorbs ultraviolet light such as that emitted by the sun. In 1913, a French physicist, Charles Fabry, using a state-of-the-art instrument he had developed, called an "interferometer", found that ozone, in addition to being present near ground level, was also present high in the stratosphere with a concentration of about eight molecules per million molecules of air. The ozone layer lies between 15 and 35 km (9 and 22 mi.) altitude, with peak concentrations at about 25 km (16 mi.). It varies in height from 25 km (16 mi.) at the equator sloping down to 15 km (9 mi.) at the poles. The concentration at mid-latitudes is about 10^{18} to 10^{19} molecules per cubic meter; the concentration at high latitudes is greater than at low latitudes.

The presence of stratospheric ozone is vitally important. First, the process of ultraviolet light absorption heats the air in the stratosphere. The temperature increases with altitude, and provides a stable thermal structure in the stratosphere. Second, the ozone layer protects life from the very damaging ultraviolet fraction of the sun's radiation. This shielding action is essential, for without it life (including of course human life) would never have evolved. If compressed, the ozone layer would only be as thick as a wafer. This is all that stands between us and extinction.

THE FORMATION OF OZONE

A Briton, Sydney Chapman—one of the greatest geophysicists of all time and founder of the International Geophysical Year of 1957–1958—first explained how ozone is formed in the upper atmosphere. When sunlight of a wavelength less than 240 nanometers (one billionth of a meter) is absorbed by an oxygen molecule, the energy of the light splits the oxygen molecule into two oxygen atoms. The atoms then continue to bounce around until one strikes and joins a second oxygen molecule to create an ozone molecule, O_3. Ozone is relatively unstable, and is broken up rapidly by sunlight of a wavelength less than 1,100 nanometers, as well as by colliding with other atoms. Although ozone molecules are constantly being made and destroyed, there are always some present in a relatively constant concentration (Emiliani, 1991; Cagin and Dray, 1993).

Ultraviolet Radiation

To appreciate the importance of ultraviolet (UV) radiation, we must understand how it interacts with living cells. An important property of radiation is that as its wavelength decreases, its energy increases. In other words, short wavelength radiation (like ultraviolet light) has more energy than long wavelength radiation (like radio waves). UV radiation has a wavelength of about 200–400 nanometers and is divided into three ranges:

- UV-A rays with wavelengths of 320–400 nanometers are the weakest and generally harmless to humans (although they accelerate aging and can cause wrinkles).
- UV-B rays with wavelengths of 280–320 nanometers are harmful to humans and the environment. The shorter wavelength components can penetrate human skin and are sufficiently energetic to break and tangle the bonds within DNA molecules, which may lead to cancer (Cagin and Dray, 1993). Although the DNA damage occurs instantly, the ensuing cancer may not take place for many years. UV-B also causes cataracts, suppression of the immune system, and changes to ecosystems.
- UV-C radiation with wavelengths of 200–280 nanometers is very harmful and even a small exposure would be lethal to humans, animals, and plants (Fisher, 1990). Fortunately, the atmosphere absorbs all the UV-C radiation.

Eons ago, when life first started to evolve, the ozone layer had not yet formed. For this reason, life did not begin on land; early life evolved in the oceans, where a shield of about 10 m (33 ft) of water protected it from UV radiation.

UV radiation has a wavelength of similar size as cellular structure and, combined with its high energy, can interact with cells and cause damage. Because of its deadly power, UV radiation is commonly used as an antiseptic, a bleaching agent, and a drinking-water purifier. In contrast, longer waves—such as shortwave radio waves, with wavelengths of several meters— pass right through living matter without causing any problems.

ELECTROMAGNETIC RADIATION

UV radiation is only one small part of the spectrum of electromagnetic radiation. The most familiar part of this spectrum is visible light, which—when viewed with a prism—consists of a rainbow of colors. Each color is a specific wavelength ranging from red (with a wavelength of about 700 nanometers) to violet (with a wavelength of about 450 nanometers). Although we cannot see it, we are also surrounded by radiation of wavelengths both longer and shorter than visible light. Radiation with increasingly longer wavelengths includes infrared radiation, television and radio waves, and shortwave radio waves. Radios, for example, receive waves with a length of about 1 meter (1.1 yards) that are in the air all the time, whether the radio is turned on or not. Radiation with wavelengths shorter than visible light include, in order of decreasing wavelength, UV radiation, X-rays, gamma rays, and cosmic rays. Because the energy of radiation increases as its wavelength gets smaller, those rays that are shorter than visible light are very energetic and can be damaging to health. This is why, for example, your dentist and doctor keep the number of X-rays they take to a minimum.

Ultraviolet light is a serious health concern. Over 130,000 cases of melanoma skin cancer are recorded each year around the world. In 2004, about 20 million people worldwide were blind from cataracts. According to the World Health Organization, up to 20 percent of these cases may be caused or exacerbated by sun exposure. In 1991, the United Nations Environmental Program predicted that a long-term 10 percent depletion of the ozone layer would cause an additional 300,000 cases of skin cancer per year worldwide. It would also cause 1 million additional cases of cataracts per year worldwide. Increased UV exposure also causes premature skin aging, increased sunburn, and suppression of the immune system leading to an increase in infectious diseases and a decrease in the effectiveness of vaccinations. Similar effects are expected for animals; although fur and feathers protect them to some degree, their eyes and faces are vulnerable.

We will see shortly how the development of the ozone holes over the poles is a seasonal phenomenon. When the Antarctic ozone hole breaks up in November, sometimes large pieces break off like bubbles and float north, creating risks to more populated areas in the southern hemisphere. This problem could be even worse in the northern hemisphere. In Queensland, Australia, more than 75 percent of people 65 years and older have some form of skin cancer. Children are required by law to wear hats and scarves when going to and from school. In Ushuaia, Argentina, residents are warned to stay indoors as much as possible during September and October (Gore, 1992). In Tasmania, the Australian state closest to the Antarctic ozone hole, the incidence of melanomas doubled in the decade from 1980 to 1990. In Punta Arenas, Chile, the southernmost city in the world, property values plummeted by nearly one-third within a few years of the discovery of the ozone hole (Cagin and Dray, 1993), and since 1994 skin cancer rates have soared by 66 percent (Flannery, 2005).

Plants are also vulnerable. Studies have shown that increased UV radiation causes a reduction in plant height and leaf area, which leads to decreased yield of crops such as wheat, rice, corn, and soybeans. Forest growth is also stunted.

Synthetic plastic materials degrade more rapidly (Environment Canada, 1990) becoming more brittle and with their colors fading.

Increased UV radiation can also damage aquatic communities. One laboratory experiment showed that a 20 percent increase in ultraviolet light for 15 days resulted in the death of all exposed anchovy larvae (Cagin and Dray, 1993). Major destruction of the microscopic phytoplankton that live near the surface of the oceans could have devastating effects given that they form the basis for the entire marine food chain and also play a major role in regulating the composition of the atmosphere, by removing large quantities of carbon dioxide and generating oxygen (Gribbin, 1986).

The effects of ozone depletion can be reduced if people are aware of the dangers of UV radiation and take steps to protect themselves. A useful tool for informing the public about daily exposure levels is a UV Index.

UV RADIATION A DINOSAUR KILLER?

One theory for the extinction of dinosaurs 65 million years ago blames a temporary but severe depletion of the ozone layer. The theory holds that an unusually high period of activity on the sun, including large flares, led to an outburst of energetic solar particles. On entering Earth's atmosphere these particles interacted with the stratosphere to create large amounts of active nitrogen compounds that are potent ozone depleters. The ozone layer was reduced substantially for a short time, leading to the tiny UV rays killing the enormous dinosaurs. This theory, however, does not receive wide support simply because all other life would also have been damaged, and there is little evidence for such mass destruction (Cagin and Dray, 1993).

In 1992, Canada became the first country to issue a UV Index in daily weather forecasts.

The Canadian UV/ozone observational network consists of 12 stations, of which three are located in the Arctic. Four stations have data dating back to 1957; observations at the rest began in the late 1980s and in the early 1990s. Readings are taken every five minutes and transmitted to a supercomputer in Montreal, which calculates the thickness of the ozone layer and factors in the angle of the sun's rays, which varies with the seasons. It incorporates the effects of cloud cover and weather to issue daily UV Index forecasts for the entire country.

The World Health Organization recognized the value of the UV Index, particularly for tropical countries where the UV levels are naturally high, and together with the World Meteorological Organization in 1994 developed a global UV Index based on the Canadian model. This has now been adopted as an international standard and is used in over 26 countries. The index ranges from zero to greater than ten. The range between zero and four is considered low; four to seven is considered moderate. The range above seven is high. Values greater than nine are seldom seen in Canada, whereas values above 13 can be recorded at tropical beaches.

CFCs: "Wonder Chemicals"

As late as 1970, scientists had no idea that the ozone layer was being depleted. And once the ozone hole over the Antarctic was discovered, it took more time and considerable detective work before the culprit was discovered. Much to the surprise of the scientific community, the damage was being caused by synthetic compounds that were once hailed as "wonder chemicals".

In 1928, Thomas Midgley Jr., an engineer at General Motors' research laboratory in Detroit (whom we met in Chapter 5), was looking for a way to provide cooling for car interiors. Refrigerants in those days were either flammable or toxic, or both—so that cooling was rather primitive, and most often involved ice. Midgley examined the periodic table, looking for elements whose boiling points were suitably low for use as a refrigerant. Dismissing those elements that were too unstable, he observed that fluorine remained as a potential candidate. Midgley thought that fluorine's extreme toxicity and corrosiveness could be negated by binding it into a compound so strongly that it could not escape and react with other chemicals.

After numerous experiments he discovered two gases, trichloromonofluoromethane (CCl_3Fl) and dichlorodifluoromethane (CCl_2Fl_2), which fit the bill. He called them "chlorofluorocarbons" or CFCs. The former is also known as CFC-11 and the latter as CFC-12. These molecules have a similar structure to methane (CH_4; four hydrogen atoms attached to one carbon atom) but with the hydrogen atoms replaced by chlorine and fluorine (see Figure 8-1).

FIGURE 8-1

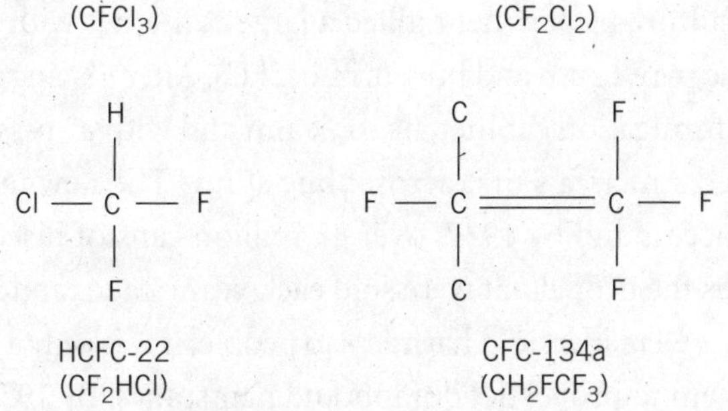

CFC-11
($CFCl_3$)

CFC-12
(CF_2Cl_2)

HCFC-22
(CF_2HCl)

CFC-134a
(CH_2FCF_3)

Chemical structure of CFCs

CFCs were thought to be ideal chemicals because they are extremely stable, nontoxic, and nonflammable and so would not cause any environmental or health problems. In addition, they are very compressible and make excellent refrigerants and propellants. In 1930, Midgley gave a sensational demonstration of the safety of his new refrigerants. Into a glass dish he poured liquid CFCs, which on contact with room temperature air started to boil. Dramatically, Midgley leaned over the dish and deeply inhaled the rising vapors. Then he slowly turned to a burning candle on the table and exhaled, extinguishing the flame. "This refrigerant", he explained, "is non-explosive and we believe non-poisonous. It has not harmed animals. I have breathed quantities of it without lasting bad effects" (Cagin and Dray, 1993).

By 1938 the new compounds, marketed by DuPont under the name Freon, had captured 15 percent of the refrigerant market and their use was growing rapidly. Soon other companies were also manufacturing these compounds, which became the wonder chemicals of the twentieth century. Midgley's invention of CFCs transformed society and within a few decades air conditioning was a way of life, used in cinemas, restaurants, factories, offices, cars, and homes. By 1971, air-conditioning units were installed in 58 percent of the 9.3 million cars in use. It is fair to say that CFCs made possible the urbanizing and flourishing of large regions of the southern United States.

And the apparent wonders of CFCs continued. Malaria was a devastating disease feared around the world. Fumigation, which received an enormous boost from the research into gas warfare during World War I, was the main method of eradicating mosquitoes, the carriers of malaria. The application of insecticides, however, was cumbersome, requiring heavy canisters with liquid spray pumps. In 1941, Lyle Goodhue, a chemist in the US Agriculture Department, filled a large container with a mixture of the insecticide pyrethrum and pressurized CFCs. After placing several cockroaches in a fumigation cabinet, he took aim and with a "psssst", successfully deployed America's first aerosol bug spray. The new invention was an instant success and by 1947, over 45 million cans of insecticide spray using CFCs as the propellant were sold each year (Cagin and Dray, 1993).

Since CFCs were inert and harmless to people, it was only a small step to start using them to propel deodorants and hair sprays. By 1958 more than 500 million aerosol cans were being sold annually, and by 1968 the number had reached 2.3 billion cans per year. The use of CFCs doubled every five to seven years from 1950 to 1974 and peaked in the late 1970s. CFCs were also used in smaller quantities as solvents in degreasers and cleaners, to dilute sterilizing gas mixtures, and in foam packaging and insulation.

No class of chemicals has ever brought so many people comfort. Needless to say, satisfying this prodigious demand required an enormous quantity of CFCs.

A Stratospheric Problem

The first inkling that CFCs might be a problem came in 1970. A British scientist, James Lovelock, working from a laboratory in his home and using an instrument of his own design that could measure minute concentrations of gases, found traces of CFCs in the stratosphere. In fact, his preliminary calculations showed that the atmospheric load was approximately equal to the entire quantity of CFCs produced worldwide up to that time, as might be expected given their inherent stability (Fisher, 1990). He concluded that eventually all CFCs released at ground level would drift up to the stratosphere (although it might take six years or longer).

JAMES LOVELOCK

James Lovelock is one of the main ideological leaders in the development of environmental awareness. Born in 1919, he graduated as a chemist and then received PhD degrees in both medicine and biophysics. Lovelock's scientific work was distributed almost equally among medicine, biology, and instrument science. His electron-capture detector revealed for the first time the ubiquitous presence of pesticides in the environment, information that enabled Rachel Carson to write her 1962 book *Silent Spring*—the publication that played a major role in initiating the environmental movement. Later, Lovelock's invention helped to detect the presence of polychlorinated biphenyls (PCBs) in the environment. In 1970, the detector was responsible for the discovery of nitrous oxide and CFCs in the stratosphere.

Lovelock is also the author of *The Gaia Theory*, which considers Earth to be a self-regulating living being, and was named by *Prospect* magazine as one of the world's top 100 public intellectuals (Lovelock, 2006).

FIGURE 8-2

James Lovelock, inventor of the Gaia concept and first to detect CFCs in the stratosphere (Bruno Comby; www.ecolo.org)

In 1973, Mario Molina, a soft-spoken, slightly built man born in Mexico City, began his postdoctoral studies with F. Sherwood Rowland, a professor at the University of California at Irvine. Spurred by the findings of Lovelock, Rowland suggested Molina investigate the ultimate fate of CFCs; even though they were very stable in the lower troposphere, this might not be the case at high altitude under the full blast of radiation from the sun.

They soon found that CFCs were in fact unstable in the stratosphere, breaking apart under strong UV radiation. To make matters worse, the process released chlorine atoms. It was clear—since both ozone and chlorine atoms are highly reactive—that the two would interact, resulting in ozone destruction. The reaction is denoted as follows:

$$Cl + O_3 \longrightarrow O_2 + ClO$$

The most startling finding, however, was that chlorine monoxide, ClO, was also unstable and would soon break apart, freeing the chlorine atom to again react with an ozone molecule. So a single chlorine atom could convert not just one, but hundreds of thousands of ozone molecules into oxygen. This revealed that the ozone layer was indeed in real danger. Rowland and Molina calculated that the ozone layer could decrease by about 20–40 percent within a century. In a classic piece of understatement, Molina later said, "It seemed like bad manners for men to put a chemical into the atmosphere without knowing exactly what happens to it".

A NOBEL SCIENTIST WITH A SOCIAL CONSCIENCE

Born in 1927 in Delaware, Ohio, F. Sherwood Rowland earned his undergraduate degree in chemistry at his hometown Ohio Wesleyan University. A broad-shouldered man standing 1.96 meters tall (six ft, five in.), he was a basketball star in university. By coincidence he grew up a scant 22 km (14 mi.) from the home of Thomas Midgley, Jr., the inventor of CFCs. Rowland received a PhD in radiochemistry in 1952 at the University of Chicago. In 1964, he became chair of the new department of chemistry at the University of California at Irvine. From early in his career he felt that scientists should have a social conscience and took an interest in environmental issues. For example, he studied mercury content in tuna off the California coast. Not only did he and Molina make the key discovery that started the "ozone wars", but he was a keen participant in those battles. In 1995, Rowland, Molina, and Paul Crutzen were awarded the Nobel Prize in chemistry for their work.

The publication of their results in *Nature* magazine in 1974 was soon picked up by the media and headlines blared a coming doomsday. "Tests show aerosol gases may pose threat to earth", headlined *The New York Times*. "Not with a bang but with a psssst! Do aerosol cans spell environmental doomsday?" shouted the *Christian Science Monitor*. These headlines heralded the beginning of an epic, lengthy battle to ban CFCs, in which Rowland became deeply involved. As he explained, "We realized there were no other spokesmen. As soon as that became clear, we never questioned the need to go public ... we had a responsibility to go public" (Cagin and Dray, 1993). The battle to save the ozone layer dragged on for 13 years but in the end it resulted in international cooperation to drastically curtail the use of CFCs, through the signing of the Montreal Protocol in 1987.

SUPERSONIC AIRPLANES AND OZONE

Paul Crutzen at the Max Planck Institute for Chemistry in Germany showed that nitrogen oxides (NOx) also destroy ozone, which led to concerns about supersonic planes, whose engines emit those compounds. In the 1960s, several countries were envisioning large fleets of supersonic airplanes that would crisscross the globe in stratospheric flight. Harold Johnston at the University of California at Berkeley calculated that the NOx emitted by such a fleet operating for two years would deplete the ozone layer by at least 10 percent and perhaps as much as 90 percent (Fisher, 1990). Although economics was the major factor for abandoning the supersonic transport program by the United States and other countries, concern for the ozone layer was also an important factor. With the demise of the Concorde program in 2003, the ozone layer is no longer damaged by commercial supersonic flights.

The evidence against CFCs continued to mount. Studies into atmospheric damage caused by rockets showed independently that chlorine destroys ozone and that the chlorine atoms remain free to continue destroying more ozone molecules. But as rockets do not generate large quantities of chlorine, the significance of this work was not appreciated initially.

Although CFCs are the main culprits, they are aided and abetted by other ozone-depleting substances including halons, carbon tetrachloride, methyl chloroform, and hydrobromofluorocarbons. These compounds, collectively known as "halocarbons", all contain either or both of the two deadly atoms chlorine and bromine (bromine is very similar to chlorine in chemical properties). Methyl bromide, MeBr, was sometimes used for fumigating soil on farms, but its use was phased out in both the United States and Canada in 2001. Halons are used in fire-extinguishing systems for electronic equipment (for example, in computer rooms) because they leave no solid residue and are safe to people. However, they are ten times more effective in destroying ozone than CFCs. Dust and acid in the stratosphere, which can be created by volcanic eruptions, also deplete ozone (Johnson, 1993).

In 1975, the concern about CFCs increased when it was found that CFCs are also potent greenhouse gases and contribute to global warming. Calculations showed that at the then-current production rate of CFCs, the global temperature could rise by an average of 0.9 C° (1.6 F°) by 2000,

NUCLEAR HEADACHE

It has been shown that nuclear bombs detonated during a nuclear war could destroy 30–70 percent of the ozone layer (McKibben, 1989). This would have an effect more devastating than the immediate impact of the bombs themselves—as if there wasn't enough to worry about in the event of a nuclear holocaust!

a significant increase that would cause disruptive climate changes (Cagin and Dray, 1993).

The "Ozone Wars"

By the mid-1970s, the enormous danger to life on the planet had been recognized and appropriate actions should have been taken immediately. But CFCs had become too solidly entrenched in modern life and formed the basis of an enormous industrial juggernaut. Instead of quickly moving to rectify the situation, the industrial sector fought a strong rearguard battle that dragged on for more than a decade. Meanwhile, the ozone layer continued to erode.

KNEE-JERK DENIAL

Denial by those with entrenched positions is not unique to the "ozone wars". It is almost an axiom that the industries responsible for environmental or health damage will claim there is no real problem, that more research is necessary, and that solutions will cause enormous economic suffering. Similar tactics were used to deny that cigarette smoking causes cancer, that lead in gasoline is toxic, and that coal-burning power plants cause acid rain. But never have we witnessed a campaign as frenzied as the one currently being waged to deny global warming.

A major debate began at the international level. At that time about 50 percent of CFCs was used in aerosol cans. In spite of strong industrial opposition, legislation was finally passed in the United States in 1976 requiring CFC aerosols in cans to be labeled as toxic. In 1978, the United States banned the use of CFCs in aerosol cans altogether and soon other countries did likewise. Air conditioning and cooling units in buildings, cars, and refrigerators became a serious problem—there were occasional leaks—and as they were taken out of service the CFCs eventually escaped.

The industrial producers led by DuPont, however, lobbied hard and effectively against any further restrictions in the use of CFCs. In the early 1980s, the United Nations tried to ban CFCs, but with strong opposition by the CFC industry, had no success. The only agreement was that more research was needed.

CHEMICAL MOTTOES

There is a certain irony in the motto used from 1935 to the 1980s by DuPont (one of the main distributors of CFCs): *Better things for better living... through chemistry*. By 1970 the image of chemistry, especially in the manufacture of synthetic chemicals, was becoming tarnished. The general public began to understand the role of CFCs in creating the hole in the ozone. Also, largely as a result of Rachel Carson's book *Silent Spring,* as well as a growing environmental movement, people became aware of the dangers of pesticides. So in the 1980s DuPont strategically dropped the tag line "through chemistry". In 1999, the DuPont trademark motto became, "*The miracles of Science®*".

In 1926, Gordon Dobson, a physicist at the University of Oxford, designed what became known as the "Dobson spectrophotometer" that more than half a century later was to play a key role in the discovery of the ozone hole. Using a quartz prism that splits sunlight into its component wavelengths, the instrument measures the quantity of ozone in the atmosphere and expresses the results in Dobson units (DU). As the instrument is robust and simple to use it became the standard for measuring ozone from the ground and was still in common use in the early 1980s. By this time, there were 70 Dobson instrument stations in the world that tracked daily and monthly ozone levels in the stratosphere.

Joe Farman of the British Antarctic Survey operated one of these stations at Halley Bay, a remote outpost in Antarctica; by the early 1980s he had been quietly recording Dobson measurements for almost 25 years. Since 1977 he had measured a slight decrease in ozone levels during the southern hemisphere's spring. However, in October 1981 his Dobson meter registered ozone levels about 20 percent below the 300 DU he was accustomed to

observing—an enormous change. Thinking he had an instrument malfunc-
tion he ordered a new measuring device, only to read the same results.
Farman, predicting the uproar that would follow the publication of these
results, wanted to be absolutely sure of his data. He repeated the measurements the following spring at another station about 1,600 km (990 mi.) away
from the first station. Again he recorded the same depleted-ozone results.
He checked and rechecked until finally in 1985, with two colleagues, he published the results. The study indicated, conclusively, an ozone depletion of
30–40 percent over Antarctica (Farman et al., 1985).

Farman, who had worked in quiet anonymity all his life, now found
himself in the middle of a raging controversy. Many scientists were skeptical, given that a far more sophisticated project had been in progress since
1978. High in the sky a US satellite called the Nimbus 7—which looked like
an ocean buoy—circled the Earth in a polar orbit every 104 minutes. Aboard
the Nimbus 7 were two instruments called TOMS (Total Ozone Mapping
Spectrometer) and SBUV (Solar Backscatter UltraViolet instrument). These
instruments were designed to measure ozone and ultraviolet light, respectively; the results were fed into computers that recorded and managed the
enormous amounts of data. This high-technology research showed no ozone
depletion over Antarctica (Cagin and Dray, 1993).

One can imagine the red faces at the NASA Goddard Space Flight
Center, home of the Nimbus program, when they learned of the British
findings. They immediately checked their own measurements. Amazingly,
Nimbus had recorded a depleted ozone layer, but its computers had been
programmed to ignore readings that greatly deviated from the norm.
Fortunately, however, all the data had been saved and could be re-exam-
ined. It was particularly embarrassing for the Americans to be scooped
by the British—who had operated on a shoestring using an inexpensive
instrument from the 1930s. Farman's annual budget of $18,000 was small
change for the multi-million-dollar American project.

A Japanese team at their Antarctic research station had already seen
low ozone values, and unlike the British, had presented them at an international conference late in the summer of 1984. However, their surprising results were not taken seriously and it was another year before Farman's
results confirmed the Japanese findings. The combined findings of the
three nations galvanized unprecedented global action. By 1986, the media
were reporting the situation as a "crisis" and the entire world grieved that

isolated Antarctica with its vast snowy plateaus, mountain ranges, glaciers, and its unusual biological life should so unfairly bear the brunt for
the consumer profligacy of the populated continents.

Even so, the $8-billion-per-year CFC lobby argued that the hole in the
ozone layer was only a temporary abnormality (Kidd and Kidd, 1998).
"Industry always said that we'd have plenty of advance warning of any ozone
problems," Rowland said in frustration in 1986, "but now we've got a hole
in the atmosphere that you could see from Mars" (Cagin and Dray, 1993).

In 1987, a major expedition with more than 150 scientists representing
19 research organizations from four countries converged on Antarctica. To
collect their data they used ground stations, balloons, satellites, and two
airplanes (including the special high-altitude Lockheed ER-2, a direct descen-
dant of the legendary U-2 spy plane of Cold War fame). The ER-2, almost
as light as a glider, was preferred for research to more modern planes because
it was the only one that could reach the stratosphere and fly at subsonic
speed, enabling the collection of air samples. This plane carried only one
person—who very likely felt some qualms about flying all alone in the harsh-
est weather conditions on the planet. The other plane was a DC8; although
it could not fly as high, it could carry several scientists and a much larger
payload of scientific instruments. This expedition confirmed Farman's
results and provided further insight into the cause of the depletion.

Since then, numerous missions and experiments have conclusively
demonstrated that ozone had been severely depleted in the Antarctic at altitudes between 12 and 20 km (7.5 and 12 mi.), and that the concentrations
of chlorine monoxide were 300 times greater than those usually observed
at these stratospheric altitudes. In fact, measurements showed the ozone
layer was depleted by about 35 percent with some areas seeing a depletion
of 70 percent. One pocket even revealed a depletion of 90 percent.

Earth's ozone layer continued to grow thinner, not only over the poles,
but everywhere else as well (Johnson, 1993). From 1985 to 1993, there
was a 30 percent increase of chlorine in the overall ozone layer. In southern Canada, for example, the ozone layer thinned about 7 percent from
the 1980s to the late 1990s. In February 1989, severe depletion was discovered over the Arctic; by the spring of 1997, ozone levels were down 45
percent from normal. In 1992, high-altitude balloon missions over
Antarctica and NASA satellite measurements observed that the ozone hole
was larger than ever. It was nearly three times the area of Canada and

about 25 percent larger than ever previously measured. Of particular concern was the discovery that the edge of the hole now hovered over a populated island near the tip of South America (Phillips, 1993).

WHY IS OZONE LOST MOSTLY AT THE POLES?

It took scientists some time to realize that the unique weather at the poles greatly aids the destruction of ozone (Johnson, 1993). As winter arrives, a strong vortex of winds develops around the pole, isolating the polar stratosphere. In the absence of sunlight, air within the vortex becomes extremely cold. At temperatures of −80°C (−112°F) or lower, clouds of ice, nitric acid, and sulfuric acid begin to form at altitudes of about 10–20 km (6.2–12.4 mi.), the same height as the ozone layer. This phenomenon, called "polar stratospheric clouds", has been observed since the early days of Antarctic exploration. Due to the extreme height of these clouds, someone at ground level will see the clouds reflecting sunlight long after the sun has set. The clouds are especially noticeable because they are accompanied by unusual light displays. Inactive forms of chlorine (notably hydrochloric acid, HCl), are drawn into the polar stratospheric clouds where, due to a phenomenon called "surface catalysis", chemical reactions on the surface of the ice crystals speed up the release of chlorine molecules. The molecules stick, apparently harmlessly, to the surface of the ice until spring, when then the ice melts. At this point, light breaks the chlorine molecules into two chlorine atoms that destroy ozone. Over the Antarctic these processes commonly lead to the formation of a massive ozone hole. By early November, the Antarctic polar vortex breaks up and regular air from outside mixes with the "hole". The cycle begins again the following winter.

This process illustrates the enormous complexity of the atmosphere. We have much to learn about the physical and chemical processes that take place high above our heads.

Canada and the Arctic

Canada, with a large political and economic stake in the polar Arctic region, is making a major contribution to the understanding of the ozone problem. A new ozone-measuring instrument, the "Brewer spectrophotometer", was invented in 1982 by three Canadians—David Wardle, Jim Kerr, and Tom McElroy—and since then has slowly been replacing the Dobson instrument as the international standard. By reading light from the sky, sun, or moon, it measures the wavelengths absorbed by ozone and a few other molecules. It can measure total ozone to 1 percent accuracy and also detects ultraviolet light, which allows calculation of the UV Index (also developed by Wardle, Kerr, and McElroy). About 165 Brewer instruments are now used in 43 countries.

The High Arctic Ozone Observatory was established on Ellesmere Island in 1993 and operated until 2001, allowing scientists from Canada, Japan, and the United States to conduct research and take ozone readings using a Brewer spectrophotometer as well as a number of other instruments as part of the international Network for the Detection of Stratospheric Change. High-altitude balloons are also used to measure ozone loss over the Arctic.

In 2003, the Canadian SCISAT satellite was launched by NASA into an orbit 650 km (400 mi.) high. Its main purpose is to understand the chemistry of ozone-depleting processes, especially over Canada and the Arctic. The payload includes a "Fourier-transform spectrophotometer", which measures the intensity of light in the infrared range and allows the densities of many chemicals to be determined. A second instrument named MAESTRO (Measurement of Aerosol Extinction in the Stratosphere and Troposphere Retrieved by Occultation) uses visible and ultraviolet light to provide high-resolution data on ozone concentrations throughout the atmosphere. Together the two instruments allow accurate estimates of both chemical loss and ozone movement in the polar winter and spring.

Researchers have found that quantities of ozone over the Arctic have not fallen to levels as low as over Antarctica. This is partly because the Arctic has more ozone to start with, but also because atmospheric circulation in the northern hemisphere is more variable; as a result, the stratosphere over the Arctic is warmer than it is over Antarctica. Although a polar vortex forms in the Arctic winter, it is weaker, less symmetrical, and breaks up earlier than the one at the South Pole. The Arctic ozone layer, however, is continuing to thin. In 1997 the level of ozone depletion was one of the worst on record with a recorded depletion of over 40 percent directly over the North Pole (Meteorological Services Canada, 2003). Scientists suspect that in addition to halocarbons, other factors such as climate change are changing the Arctic atmosphere and contributing to accelerated ozone depletion.

Many nations monitor the ozone layer. In 1976 the United Nations World Meteorological Organization launched the Global Ozone Monitoring program. Today, over 140 ground-based stations, supplemented by satellites, form an international ozone-monitoring network and provide regular updates on how the ozone layer is evolving.

A Solution, Finally

After much time and considerable conflict, a resolution to the ozone problem was finally reached. Here is how the story unfolds.

In 1977, the United Nations Environmental Program was sufficiently shaken by the discoveries of Rowland, Molina, and other scientists that it established a committee on the ozone layer. In the same year Scandinavia, the United States, and Canada banned the use of CFCs in aerosol cans. In 1985, the Vienna Convention on the Protection of the Ozone Layer was held, but reached no consensus other than the need for further study.

Then the news from the Antarctic expeditions broke and the consensus rapidly developed that action was urgently needed. On September 16, 1987, the Montreal Protocol on Substances that Deplete the Ozone Layer was signed by 24 countries, with the commitment to implement actual controls (including reducing CFC consumption to 50 percent of 1986 levels by 1999). The protocol allowed time for an orderly reduction of the use of CFCs and for the development of substitute products. Developing countries were permitted to increase their annual production by up to 10 percent per year for the next ten years, a concession that was seen to be necessary for countries like India and China where the widespread use of refrigeration was an important short-term goal. The Montreal Protocol, the first truly international effort to protect the environment, was a milestone (Environment Canada, 1999).

Recognizing that international opinion had clearly turned against CFCs, in 1988 the major industrial manufacturers of CFCs in the developed countries agreed to phase out their production.

As the realization came that ozone depletion was not confined to Antarctica, the international community took even more drastic steps. In 1990, the London Agreement was signed by 93 countries, including India and China, which accelerated the schedules adopted at Montreal. It required a total phase-out of CFCs by 2000 for developed countries, and by 2010 for less-developed countries. In addition, a committee was established to administer funds to poorer nations to assist them in developing alternative technologies. By 1999 there were 164 signatories and annual meetings were scheduled to review progress and make additional updates.

POLLUTION SHIELD

At mid-latitudes the effects of the high-level ozone thinning are not as damaging as had been predicted, given the extent of lost stratospheric ozone. It was concluded that something in the atmosphere is reducing the amount of solar energy reaching the ground surface. That something appears to be pollution, or smog, at ground level. We are replacing, at least to a small degree, the ozone shield with a pollution shield (which, ironically, also contains ozone) (Young, 1990).

Is the international effort working? It appears that it is. In the 1980s, DuPont introduced the new refrigerant HCFC-22 (hydrochlorofluorocarbon); its structure is shown in Figure 8-1. Because HCFC-22 has one hydrogen atom in the molecule, it is less stable and breaks up before most of it reaches the stratosphere. This new refrigerant, HCFC-22, is only 5 percent as damaging to the ozone layer as the original CFCs.

In 1993, DuPont introduced hydrofluorocarbons (HFCs), a series of new ozone-friendly refrigerant gases. Although these gases do not destroy ozone, they are not without complications. Although HFCs help prevent one environmental problem, they are potent greenhouse gases, and so contribute to another problem. HFC-134a has smaller molecules than CFC-12, so it will leak out of older air-conditioning systems unless all of the hoses and some valves are replaced with tighter, more secure parts. Ultimately, further research is required to develop chemicals that are both ozone-friendly and do not contribute to warming of the atmosphere; ideally, they would have no other detrimental side effects.

Many other alternative refrigerant gases have been synthesized. Table 8-1 lists some common halocarbons and their properties, including their ozone-depleting potential (ODP) and global-warming potential (GWP).

TABLE 8-1: Properties of Selected Halocarbons (EPA, 2008b)

Chlorofluorocarbons (CFCs)

Compound	Name	Formula	ODP*	GWP**	Use
CFC-11	Trichlorofluoromethane	$CFCl_3$	1.0	4,750	Chillers, blowing agent
CFC-12	Dichlorofluoromethane	CF_2Cl_2	1.0	10,890	Chillers, car air conditioners, blowing agent
CFC-113	Freon	$C_2F_3Cl_3$	1.0	6,130	Solvent
CFC-114		$C_2F_4Cl_2$	0.94	10,040	Solvent
CFC-115		C_2F_5Cl	0.44	7,370	Solvent, refrigerant

Hydrochlorofluorocarbons (HCFCs)

Compound	Name	Formula	ODP*	GWP**	Use
HCFC-22	Chlorodifluoromethane	CF_2HCl	0.05	1,350	Residential air conditioners
HCFC-123	Dichlorotrifluoroethane	$C_2F_3HCl_2$	0.02	76	Refrigerant
HCFC-124	Monochlorotetrafluoro-ethane	C_2F_4HCl	0.02	599	Sterilant
HCFC-141b	Dichlorofluoroethane	$C_2H_3FCl_2$	0.12	713	CFC replacement
HCFC-142b	Monochlorodifluoro-ethane	$C_2H_3F_2Cl$	0.07	2,270	CFC replacement

Hydrofluorocarbons (HFCs)

Compound	Name	Formula	ODP*	GWP**	Use
HFC-23	Fluoroform	CHF_3	0.0	11,700	HCFC byproduct
HFC-125	Pentafluoroethane	C_2HF_5	0.0	2,800	CFC/HCFC replacement
HFC-134a	Tetrafluoroethane	CH_2FCF_3	0.0	1,300	Air conditioners, refrigeration
HFC-152a	Difluoroethane	$C_2H_4F_2$	0.0	140	Aerosol propellant
HFC-227ea	Heptafluoropropane	C_3HF_7	0.0	2,900	CFC replacement

Table 8-1: Properties of Selected Halocarbons (EPA, 2008b) *continued*

Bromofluorocarbons (Halons)

Compound	Name	Formula	ODP*	GWP**	Use
	Halon-1211	$CClF_2Br$	7.1	1,890	Fire extinguishers
	Halon-1301	CF_3Br	16	7,140	Fire extinguishers

Perfluorocarbons (PFCs)

Compound	Name	Formula	ODP*	GWP**	Use
	Perfluoromethane	CF_4	0.0	6,500	Byproduct
	Perfluoroethane	C_2F_6	0.0	9,200	Byproduct, solvent

Other Chemicals

Compound	Name	Formula	ODP*	GWP**	Use
	Carbon tetrachloride	CCl_4	0.73	1,400	CFC feedstock, solvent
	Methyl chloroform	$C_2Cl_3H_3$	0.12	146	Solvent
	Methylene chloride	CH_2Cl_2		9	Solvent
	Chloroform	$CHCl_3$		5	HCFC feedstock
	Sulfur hexafluoride	SF_6		23,900	Insulator, cover gas

Notes:

*   ODP = Ozone depleting potential, relative to CFC-11, whose ODP has been defined as 1.0

** GWP = Global warming potential, relative to CO_2, whose GWP has been defined as 1.0

The production of CFCs peaked in 1988, and by 1995 production was reduced by 76 percent. Between 1986 and 1990 the global supply of CFCs and halons declined by 31 percent, from some 1.26 million tonnes (1.39 million tons) to approximately 0.87 million tonnes (0.96 million tons). There is a time lag of several years between a decrease in CFC production and a response in the stratosphere, given the time it takes CFCs to migrate there. In 1998, the UN Environment Programme reported that by 2050 without international action, the ozone layer would have been depleted by 50 percent at midlatitudes in the northern hemisphere and by 70 percent at midlatitudes in the southern hemisphere (Sarma, 1998). Instead, in 2006 the Scientific Assessment Panel to the Montreal Protocol projected the ozone layer in Antarctica will recover and return to normal at some point between the years 2060 to 2075. The Arctic ozone layer is expected to fully recover by 2050 (WMO/UNEP, 2006).

The road ahead may contain a few unexpected problems. In 2006, for example, satellite measurements revealed the highest loss of ozone over Antarctica ever recorded: about 40 million tonnes (44 million tons), compared to 39 million tonnes (43 million tons) in 2000. The size of the ozone hole measured 28 million km^2 (17.4 sq. mi.). The extreme loss of ozone, years after ozone protection measures were implemented, was explained by unusually cold stratospheric temperatures (Province, 2006). Alarmingly, stratospheric ozone depletion is also linked to global warming in a complex manner that is not fully understood.

Important Lessons

The thinning of the ozone layer was an ominous event. It marked the first time in Earth's long history that the actions of a single species affected an entire global system. It was a signal, an enormous flare in the sky, telling us that simple, everyday actions, when carried out by billions of people, can cause damage on a scale that affects the entire planet and everything on it. It was a serious warning that human numbers may exceed what the Earth can comfortably contain.

Ozone depletion carries another vital lesson. We must exercise caution—extreme caution—with the synthetic chemicals we produce.

Incredibly, aside from the bold international moves made to control CFCs, we humans have paid little heed to these lessons. The population continues to soar and our consumptive footprint grows ever larger. What is more, we continue to produce more and more synthetic chemicals without properly testing how they will interact with the biosphere.

We may have reached the point where solutions are beyond our grasp.

9
The Globe Is Heating Up

If Earth had an operating manual, the chapter on climate would begin
with a caveat that the system has been adjusted at the factory for
optimum comfort. Do not touch the dials.
—Jeremy Leggett, 1998

The plane droned due north from Winnipeg for three long hours, crossing endless forests and thousands of frozen lakes, until we landed at
Churchill, Manitoba, a town nestled on the edge of the Arctic. I visited
this lonely outpost a few years ago because I wanted a reprieve from the
concrete cityscape. This was about as distant from "civilization" as I could
travel by commercial means.

Perhaps because of its remoteness, this harsh environment is one of the
richest wildlife areas in the world. During the summer you can swim with
schools of beluga whales in the estuary of the Churchill River; each spring
and fall you can hear the thunder of hundreds of thousands of caribou.
You can watch the sky fill with flocks of migrating birds and you can enjoy
a glass of wine under a night sky shimmering with sheets of northern lights.

But it is the largest concentration of polar bears on the globe that has
brought fame to Churchill, and earned it the title, "The Polar Bear Capital
of the World". In the late fall the bears—the mightiest carnivores on
Earth—gather along the coast as they await freeze-up so they can move
onto the ice of Hudson Bay to catch seal, their favorite food. They often
pass right through the town, where patrols keep watch and issue warnings.
Troublesome bears are captured and placed in a special polar-bear jail.

I boarded a tundra buggy—a vehicle the size of a bus mounted on huge
wheels taller than a person—to search for bears. We were soon rewarded
when a mother and two large cubs, cuddly and white, ambled by, obliv-
ious to our clicking cameras.

As we headed back to town, I watched the stark beauty of the snow-cov-
ered tundra with a sense of unease, for even this remote place is feeling the
heavy presence of humans. Scientists have shown that global warming is

causing the ice of Hudson Bay to break up earlier and earlier. Since the polar bears hunt for ringed seals from the ice, they are getting less food each year. As a result the weight of both male and female polar bears is declining and female bears are having fewer cubs. In Hudson Bay, a study by the US Geological Survey and the Canadian Wildlife Service shows the population of polar bears fell 22 percent between 1987 and 2004.

Not only is starvation taking a toll, but so is death by drowning. Researchers were startled to find that bears are being forced to swim up to 100 km (60 mi.) across open sea because the ice floes from which they feed are melting, becoming smaller and drifting farther apart due to the warming climate. By 2004 the ice cap had receded about 300 km (185 mi.) farther north than the average of two decades ago. Although bears are well adapted to short swims of 10–30 km (6–19 mi.), they are vulnerable over longer distances. In 2005, a number of polar bears were spotted swimming as far as 100 km (60 mi.) off shore. The researchers returned to the vicinity a few days later following a fierce storm and found four dead bears floating in the water. "We estimate that approximately 40 bears may have been swimming and that many of those probably drowned as a result of rough seas caused by high winds", they reported (Iredale, 2005).

As my airplane flew back toward civilization, I felt troubled. What kind of species are we? Not only do we relentlessly consume everything in our path, we find it necessary to intrude into the remotest parts of the world. Standing on the highest rung of the animal kingdom surely we should be the protector of other species, not their exterminator.

Beside a polar bear, a 1,000-year-old Douglas fir, or a blue whale, a human being is small and weak. And we are smaller still beside the two largest natural systems on Earth, the oceans and the atmosphere. The troposphere, which is only a fraction of the entire atmosphere but which gives life to the biosphere and contains all our weather, is vast, with a volume six times all the oceans together. How could things as tiny as human beings possibly change or influence something of such enormous size? It seems inconceivable.

Yet that is exactly what we are doing. The atmosphere is an immense reservoir that has always absorbed and diluted human effluent with ease. But we are burning fossil fuels at such a rate that we are modifying the chemical composition of the entire atmosphere. Never before have humans operated on such a scale.

The chemical equilibrium of the atmosphere took hundreds of millions of years to establish, all the while interacting with the oceans and their circulations, global and regional weather patterns, forests and other plants, and animals. Fiddling with the dials of such a complex, nonlinear system is foolhardy. The jury is in, and the outcome of massive injections of carbon dioxide into the atmosphere is now clear: global temperature is rising at a rate and to levels never witnessed in the history of *Homo sapiens*. As the heat rises, the weather and climate are changing, often in ways we have never experienced. And we are caught in a spiral I call the "greenhouse vortex". As the temperature increases, we turn up the air conditioning; this uses more electricity, creating more greenhouse gases, making it even hotter.

Humans are facing an environmental problem that encircles the entire planet. Such a global-scale problem requires a global-scale solution. Yet the collective resolve to fix the problem—other than perhaps in Europe—simply is not there. No Herculean efforts are being mounted to reverse the trend.

Is the Globe Really Warming?

Global warming is a very serious issue. Yet many people are not convinced that a problem exists, and if it does, they do not deem it sufficiently important to make sacrifices or changes in their lifestyles. I have a few friends—well educated and open-minded—who are not convinced global warming is a reality. There are four major objections that they, and the fossil-fuel industry, commonly raise against taking action to curtail global warming.

First, the results of global climate change are not well defined. We are daily accustomed to large swings in weather and the newspapers regularly bring us stories of tornadoes, hurricanes, heat waves, blizzards, and other catastrophic weather. And we are all familiar with even more extreme historical climate changes: the coming and going of ice ages. Against that background, the predicted temperature changes caused by global warming seem small and insignificant.

Only as recently as the 1970s scientists were predicting an ice age, given that from 1940 to 1970 world temperature had dropped by about 0.2 C° (0.36 F°). In 1974, for example, *Newsweek* ran an article "The Cooling World" that warned of ominous signs that the Earth's temperature was "cooling down ... and that the drop in temperature over the past thirty years had taken the planet about a sixth of the way toward the next Ice Age" (Fisher, 1990).

Second, global warming is a complex and poorly understood phenom-enon. There has been considerable debate about the degree of warming that might occur and its consequences. Although the predicted warming is marked by a degree of uncertainty, the conclusion is not necessarily wrong. But there are many gray areas. It is these gray areas that skeptics attack. Groups with vested interests—and there are many, including auto-mobile manufacturers, oil and gas producers, and the coal industry—muddy the waters and, aided by the media, confuse the public. In today's society, the scientific debate is viewed almost entirely through the lens of the media, motivated by the extra sales that sensationalism, emotionalism, and alarmist headlines bring—rather than by accurate, detailed reporting.

Third, the solution to global warming is not simple. There is no easy fix. The creation of carbon dioxide, the main gas causing Earth's temper-ature to creep upward, is an unavoidable consequence of burning fossil fuels. Carbon is not an unwanted contaminant like sulfur or mercury; instead, it is the primary component that burns (oxidizes) to release ther-mal energy. As each atom of carbon in coal, oil, or natural gas burns, it produces—in addition to heat—one molecule of carbon dioxide. So carbon dioxide is unavoidable if we want energy from fossil fuels. Furthermore, there are presently no cost-effective pollution control devices (such as scrubbers for sulfur dioxide) that can capture carbon dioxide from a smoke-stack. A significant reduction in carbon dioxide emissions can be achieved only by finding alternatives to fossil fuels or by dramatically changing how fossil fuels are burned. This would require radical changes in the way energy is produced, distributed, and used; a transition that would involve major shifts in industry and jobs from established sectors to new activities served by as-yet-unrealized technologies.

Fourth, and most importantly, the economies of virtually all the world's countries are based on continuing growth. And a fundamental require-ment for economic expansion is an increasing supply of energy. For this reason, some critics feel that a decrease in greenhouse gas emissions can only be achieved by curtailing economic growth and economic welfare. Powerful lobby groups have used this argument successfully, given the general public's fear of an economic recession and an erosion in their standard of living, and concern over losing their jobs.

In spite of this skepticism, there is an overwhelming consensus among the world's scientists that the globe is definitely warming as a result of human activity. To see why they have reached this conclusion and why the preceding four objections have little merit, let us have a closer look at the phenomenon of global warming.

The Greenhouse Effect

The atoms in a molecule are bound together as if attached by springs. When a molecule absorbs energy its atoms vibrate with the springs, shortening, stretching, and bending; but they can do so only in fixed patterns determined by the number and kind of atoms in the molecule.

Carbon dioxide, water vapor, and a few trace gases in the atmosphere—collectively known as greenhouse gases—share a unique property. Their molecules are arranged in such a way that they are not at all affected by the sun's radiation entering the atmosphere. These rays pass right by. But these molecules do absorb outward-bound radiation, which has a longer, infrared wavelength. This causes the molecules to increase their vibration, thereby increasing their temperature; consequently the temperature of the atmosphere increases.

The difference in wavelengths of radiation that arrive at and leave the planet is due to the large difference between the surface temperatures of the Earth and of the sun. A scientific concept, called "Wien's Law", states that a body emits radiation with a wavelength inversely proportional to its temperature.

The sun, with a surface temperature of about 6,000°C (10,832°F), emits radiation with a shorter wavelength (about 0.35–0.75 micrometers), which does not interact with water vapor and carbon dioxide molecules; in other words, it passes through these compounds. Once the sun's radiation is absorbed by the Earth, some is re-radiated outwards. The Earth, however, with an average surface temperature of about 14°C (57°F), emits radiation with a longer wavelength (10 micrometers). Rather than allowing this radiation to pass, water vapor and carbon dioxide absorb it (Drake, 2000).

These greenhouse gases thus allow incoming sunlight to pass through the atmosphere, and then, like the panes of glass in a greenhouse, they absorb the heat rather than allowing it to radiate back into space. The term "greenhouse effect" was first coined by the French mathematician Jean Baptiste Fourier in 1827. His calculations of incoming and outgoing solar radiation showed the Earth's temperature should be an ice-cold –15°C (5°F). However, by some unknown process the atmosphere was retaining heat (Flannery, 2005). It was not until 1863, when British scientist John Tyndall was able to measure the thermal properties of water vapor and carbon dioxide, that the process was understood.

The greenhouse analogy, in fact, is not exactly correct: the glass panes of a greenhouse do not absorb radiant heat in the same way that the atmosphere does. Instead they retain heat by trapping warm air inside the greenhouse and preventing cool air from entering.

The greenhouse effect has been present for most of Earth's history and maintains the atmospheric temperature at levels acceptable for life as we know it. Without the greenhouse gases, Earth's surface average temperature would be $-19°C$ ($-2°F$) instead of its actual $14°C$ ($57°F$), or fully 33 $C°$ ($59 F°$) colder. The current problem is about additional global warming caused by human (or, as it is often called, "anthropogenic") activities.

ARRHENIUS EXPLAINS ICE AGES

Around 1900, Svante Arrhenius, a Swedish physical chemist, described the unprecedented rate at which society was mining coal to fuel the Industrial Revolution as "evaporating our coal mines into the air". He was the first to investigate the possible effects of increased carbon dioxide emission into the atmosphere. In trying to explain the ice ages, he calculated that, although they are present only in small concentrations, water vapor and carbon dioxide together absorb enough infrared radiation to warm the Earth by about 33 $C°$ ($59 F°$). He also postulated that the changes in temperature that caused advances and retreats in ice ages were caused by fluctuations in the concentration of atmospheric carbon dioxide. Although not widely acknowledged at first, his theory eventually came to be accepted. He also calculated that the average global temperature would rise by about 5 $C°$ ($9 F°$) if the amount of carbon dioxide in the air doubled from its pre-industrial concentration, a figure remarkably similar to that made by modern supercomputers. In 1927 he received the Nobel Prize for chemistry.

Not all global warming is due to greenhouse gases, although they are the dominant cause. Some warming is also due to deforestation and changes in land use, both of which result from human activities. Natural causes include changes in oceanic currents and volcanic eruptions. (It is worth noting that volcanic activity releases less than 1 percent of the emissions of carbon dioxide caused by humans.)

Greenhouse Gases

Water vapor is the most important greenhouse gas, and its abundance in the atmosphere is relatively constant. It forms part of the massive water or "hydrologic" cycle, which is largely unaffected by human activities.

The main human-produced greenhouse gases and their relative contributions to global warming are: carbon dioxide, 60 percent; methane, 20 percent; halocarbons (CFCs and their replacements), 14 percent; and nitrous oxide, 6 percent (IPCC, 2001). The relative contribution of carbon dioxide will increase in the future. Similarly, the importance of methane—which molecule for molecule creates 25 times more greenhouse effect than carbon dioxide—will also increase in the future. We saw in Chapter 8 that ozone-destroying halocarbons also contribute to global warming, providing a disturbing link between these two atmospheric problems.

Carbon dioxide is a colorless, tasteless, transparent gas, which—although comprising a mere 0.038 percent of the atmosphere—is central to the climate-change issue. It is fundamental to life and is used by plants to produce oxygen by the process of photosynthesis (incorporating carbon into the plant structure). Carbon dioxide, as we have already seen, is released into the atmosphere primarily by the massive burning of fossil fuels.

In the past it was difficult to determine atmospheric carbon dioxide concentrations accurately. However, in the late 1950s Charles Keeling installed a new instrument of his invention high on the slopes of Mauna Loa in Hawaii far from industrial pollution. He began a long series of carbon dioxide measurements, now known as the "Keeling curve". These measurements show that carbon dioxide concentrations started at about 316 parts per million (ppm) in 1959 and rose continuously to more than 382 ppm by 2006. This represents an astonishing 21 percent increase in 47 years (see Figure 9-1, NOAA, 2007). The ragged annual decreases and increases in Keeling's curve reflect plant growth in the spring and decay in the fall.

Experts estimate that the carbon dioxide concentration will increase and then stabilize somewhere in the range between about 450 ppm and 600 ppm (Houghton, 1997; Gore, 1992).

Scientists have found an elegant way to study the recent history of the atmosphere by looking at ice cores from glaciers; here, key information has

FIGURE 9-1

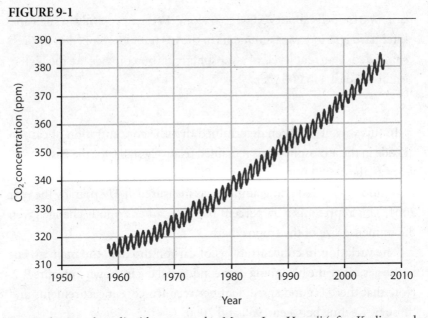

Atmospheric carbon dioxide measured at Mauna Loa, Hawaii (after Keeling and Whorf, 2004)

in effect been stored in deep freeze. Ice sheets are the archives of atmospheric history. Drill rigs are carried by helicopter to Greenland and Antarctica where they extract ice cores that can be more than 1 km (0.62 mi.) long. The cores are dated by counting annual snowfall layers. The amount of carbon dioxide in trapped air bubbles is measured, yielding atmospheric carbon dioxide concentrations from the distant past. In addition, a temperature profile can be obtained by measuring isotopes of either oxygen or hydrogen in different layers (Fisher, 1990). (Isotopes are atoms of the same element with the same number of protons but with different numbers of neutrons in the nucleus.)

EUROPEANS DRILL RECORD ICE CORE

In 2004, after eight years of tedious drilling, scientists from ten European countries extracted the last section of a core of ice 3.27 km (2.03 mi.) long from a plateau in east Antarctica. The European Project for Ice Coring in Antarctica provided information dating back 740,000 years, more than double that measured by previous cores. Analyses revealed that Earth has seen eight ice ages during this period, punctuated by brief warm spells about 25,000 years long. We are presently in such an interglacial period. Measurements of the gas bubbles trapped in the ice showed that current levels of atmospheric carbon dioxide are the highest in the past 440,000 years.

In this way, it has been determined that the concentration of carbon dioxide in the atmosphere has remained relatively stable for the past 10,000 years at about 280 ppm. Around 1800, about the start of the Industrial Revolution, it started rising, until it was measured at 377 ppm by the year 2004. This represents a 35 percent increase, an enormous change given the immense size of the troposphere.

The variation in concentrations of carbon dioxide, the main greenhouse gas, since the beginning of the industrial era is shown in Figure 9-2. Note that there is good agreement between ice-core measurements and those by Keeling between 1960 and 2000.

FIGURE 9-2

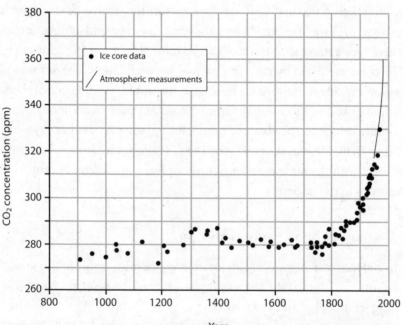

Atmospheric carbon dioxide rising in the industrial era. Circles indicate data from ice cores; after 1959 data was obtained from direct atmospheric measurement

To understand the role of carbon dioxide in the atmosphere, we must first consider its role in the planet's life cycle. Since it can bond with other atoms in a variety of different ways, carbon is the backbone of organic compounds. It is critical to all forms of life. Except for perhaps water and oxygen, there is no more important cycle in nature than the circulation of carbon between air, land, and water. The quantities of carbon involved in this global circulation scheme are enormous. The supply of carbon dioxide in the atmosphere would soon be exhausted if not for constant replenishment from the decomposition of organic matter and animal respiration. Oceans also play a key role in the carbon cycle.

The Earth has two enormous reservoirs for carbon dioxide: forests and oceans, which contain about 50 times more than the atmosphere, and have the capacity to absorb at least some of the increase of carbon dioxide in the atmosphere caused by humans. Most of the oceanic carbon dioxide remains dissolved in seawater. A small fraction is used by crustaceous sea plants and animals, which combine it with calcium to manufacture calcium carbonate to form their shells. Later these shells fall to the ocean floor where they can accumulate to large thicknesses; over time, this forms the sedimentary rock limestone. In this process, a fraction of atmospheric carbon dioxide is captured and retained for tens to hundreds of millions of years.

Until the mid-1950s, most scientists believed that the oceans would absorb all the carbon dioxide produced by humans. But oceanographers now estimate that only about 33 percent of greenhouse gas emissions are being absorbed (Schlesinger, 1997). And furthermore, the amount of carbon dioxide that the oceans can absorb is decreasing (Flannery, 2005). Keeling's observations are proof that anthropogenic carbon dioxide (emissions caused by humans) is largely remaining in the atmosphere.

Forests, particularly rain forests, take in carbon dioxide through photosynthesis and use the carbon to grow. In this way, plants remove about 90 billion tonnes (99 billion tons) of carbon dioxide from the atmosphere every year, about 14 percent of the atmosphere's total carbon content (Young, 1990). This removal is balanced by the respiration of animals, which releases carbon dioxide. But rain forests, as well as wetlands, are being destroyed while the globe's population continually grows. Land conversion, primarily deforestation in the tropics, releases about 1.5 billion tonnes (1.7 billion tons) of carbon each year, or about 0.2 percent of the atmospheric reservoir (Mitchell, 1991). In Indonesia, during some years, forest fires—most set by illegal loggers—are so vast that the smoke creates a health-damaging smog over much of southeast Asia and interferes with air and marine traffic in neighboring countries.

We are blindly creating problems at each end of the spectrum of human activity: at one end we are consuming fossil fuels that generate greenhouse gases; at the other end we are stripping away forests that have the capacity to absorb the gases.

The net result is an increase in atmospheric carbon dioxide of about 3.2 billion tonnes (3.5 billion tons) of carbon per year. Carbon dioxide emissions are expected to increase by 1.5 percent per year from 2001 to 2025 (EIA, 2003).

Methane concentration in the atmosphere, according to glacial core studies, was constant at 0.7 ppm until about 300–400 years ago, when it began to rise. Today its atmospheric concentration is 1.7 ppm and it is projected to increase at 1 percent per year between 2001 and 2025 (EIA, 2003), in parallel with human population growth. Methane will become more damaging in the future. By 2030, it could be contributing 40 percent of global warming (Mitchell, 1991).

Methane is released naturally from wetlands and by rice paddies through anaerobic bacteria. It is also emitted by landfills, coal mining, and the oil and gas industry. As human population grows so does the number of its farm animals such as cows, pigs, sheep, and buffalo. Each one of these creatures releases large quantities of methane—about 400 liters/106 US gallons per day per cow—through enteric fermentation (belching) as cellulose is broken down in the rumen (the first section of their stomachs) into digestible cud. Rather surprisingly, enormous quantities of methane are also produced by the digestive bacteria of termites whose numbers are increasing rapidly as they feed on the dead wood of destroyed forests (Fisher, 1990).

Nitrous oxide (N_2O), also known as "laughing gas", causes about 250–300 times more global warming than carbon dioxide (for the same mass) and persists in the upper atmosphere for 100–200 years (Mitchell, 1991). This trace gas has a mean concentration of 311 parts per billion (ppb); this represents an increase of about 17 percent since the Industrial Revolution. It continues to increase by about 3.9 million tonnes (4.3 million tons) of nitrogen per year, mostly due to the substantial worldwide increase in fertilizer use. Conversion of tropical forests to cultivated lands and pastures results in greater nitrous oxide emissions, and the flux increases when fertilizer and manure are applied to the lands (Schlesinger, 1997).

Ozone associated with smog in the lower atmosphere is also a greenhouse gas. It should be noted that one pollutant, sulfur dioxide when in the form of sulfate aerosols, acts with the opposite effect on the climate. While the carbon dioxide in the upper atmosphere traps heat, the sulfate aerosols at lower levels reflect sunlight back into space, creating localized cooling effects. Fossil fuels account for most of the airborne sulfate aerosols.

Yes, the Temperature Is Rising

Why are scientists convinced global warming is really taking place? First, temperatures have been recorded for over a century by many meteorological stations, and these show that the Earth has warmed by about 0.8 C° (1.44 F°) over the past 120 years. The temperature decrease from the 1930s to the 1970s turned out to be a small perturbation on a larger and longer increasing trend. By studying the oxygen isotopes of ice cores extracted from Antarctica and Greenland, scientists have determined past temperatures and have confirmed the meteorological measurements.

In 1988, the United Nations Environmental Program and the World Meteorological Organization established the Intergovernmental Panel on Climate Change (IPCC) of over 300 leading scientists from around the world. The panel was to provide an authoritative review of the subject and develop quantitative bases for strategies by which nations could reduce their greenhouse gas emissions.

In developing their global warming predictions, IPCC scientists conducted exhaustive studies and developed computer models of the climate, called "general circulation models". In these models the globe (or some part of it) is portrayed by thousands of boxes. Each box is subdivided into many layers representing the layering of the atmosphere and ocean or land. The computer program solves the fundamental conservation laws inside each box, and then calculates the transfer of energy, mass, and momentum from one box to the next throughout the model. About ten such models are in operation, with the most sophisticated located at the Hadley Centre in England, the Lawrence Livermore National Laboratory in California, the Max Planck Institute for Meteorology in Germany, and the University of Victoria in British Columbia (Flannery, 2005).

These models are highly complex; they depend on many parameters, not all of which are well understood. For example, it is not easy to mathematically model the interaction between atmospheric systems (with time scales of days and weeks) and oceanic systems (with time scales of months to years). Weather systems alone are enormously complex and include many components. Cloud cover, for example, can either cool or warm the surface below depending on its thickness, height, and the internal distribution of water vapor, ice particles, and other aerosols. The models must accurately include all these factors to be of use.

The IPCC set out to answer many questions. Will cloud cover increase with rising temperature? If yes, by how much? One hypothesis, in particular, required testing. Some scientists thought the warming trend caused by greenhouse gas emissions would be counter-balanced by natural processes such as increased absorption of carbon dioxide by the oceans, or an increase in cloud cover (which then would reflect away more sunlight). That is, they felt Earth's systems might act as a self-regulating thermostat to correct any drastic increases or decreases in temperature.

In late 1995, the IPCC concluded that human activity was a contributor to the warming of the atmosphere (IPCC, 1995). This was a major result, which clearly proclaimed that global warming could no longer be attributed to natural variations in climate. By coincidence, the World Meteorological Organization reported that the same year (1995) was the hottest year in recorded weather history.

In 2007, the IPCC issued an updated report that stated, "Most of the observed increase in global average temperatures since the mid-twentieth century is very likely due to the observed increase in anthropogenic greenhouse gas concentrations." The IPCC also stated that future changes would include: additional global warming, changing precipitation patterns and amounts, rising sea levels, and increasing frequency and intensity of some extreme weather events. It warned that the stakes associated with these climate changes are high (2007).

The 2007 IPCC report on global warming presents predictions for several scenarios. The one that matches current conditions (society relying primarily on fossil fuels) predicts a 4.0 C° (7.2 F°) increase in temperature by 2100 relative to 1990, with a range of uncertainty of 2.4–6.4 C° (4.3–11.5 F°). Another scenario, in which society uses a balance of fossil and renewable energy sources, predicts a 2.8 C° (5 F°) increase in temperature by 2100 relative to 1990, with a range of uncertainty of 1.7–4.4 C° (3–7.9 F°).

The model studies are continually being upgraded and include many factors such as the potential cooling effects of increased cloud cover, sulfate aerosols, increased rate of plant growth, and increased oceanic uptake. They showed that self-regulation by the Earth, if present, is being overridden. Some of the models' range of uncertainty is due to gaps in understanding the science of climate, but much of it is due to uncertainty in human demographics such as population growth rates and whether fossil fuels will be used less in the future. Furthermore, the panel predicted global temperature will continue to increase after 2100.

An average rise of 4.0 C° (7.2 F°) is more significant than it might appear. First, the distribution of temperature increases will not be uniform across the globe. Instead, land will warm more than oceans, and greater warming is predicted for high latitudes, and less warming near the equator. The Arctic has already warmed significantly, which is particularly important for the polar ice sheet and the Arctic habitat. In Canada, the annual mean temperature could increase by 5–10 C° (9–18 F°) over the next century.

Second, ice core samples from Greenland show that during the warming period after the most recent ice age, the average Greenland temperature rose about 8 C° (14 F°), which is equivalent to an average worldwide temperature rise of 2–3 C° (3.6–5.4 F°). Ice cores show a strong correlation between global temperature changes and carbon dioxide concentrations dating back over 400,000 years. These data show that the average global temperature has varied by as much as 10 C° (18 F°) between ice ages and interstades (warm periods), but by less than 4 C° (7.2 F°) in the past 10,000 years (UNEP, undated). A change of more than 3 C° (5.4 F°) would be the greatest during human civilization, and would have very significant effects. For example, scientists believe that at current rates all polar ice will be melted by 2100 (Grady, 1997).

The rate of temperature increase is also a concern. Ice ages have been directly linked to what are called "Milankovitch cycles", that is, the variations in the amount of solar radiation reaching Earth because of changes in: (a) the elliptical shape of the Earth's orbit around the sun—a 90,000- to 100,000-year cycle; (b) the tilt of the Earth's axis—a 40,000-year cycle; and (c) the precession (a slow movement of the axis of rotation of a spinning body about another line intersecting it) of the Earth's axis—a 21,000-year cycle. Sometimes these cycles reinforce and sometimes they oppose each other. The combined results when calculated over past millennia bear an impressive correlation to the geologic record of temperature change as determined from oxygen isotopes and fossilized marine organisms. Milankovich's work is regarded as one of the greatest contributions to climate studies.

The period from 20,000 to 10,000 years ago saw the last ice age end and brought the fastest temperature rise recorded in Earth's history; about 1 C° (1.8 F°) per thousand years. The changes we are seeing today due to the emission of greenhouse gases are about 30 times faster than those historical changes (Flannery, 2005).

Natural systems are not accustomed to dealing with rapid changes. Fossil fuels were created slowly over tens of millions of years. Now, we are ripping them out of the ground and transforming them into a greenhouse gas, and this is happening in the space of about a century—a barely perceptible heartbeat compared to the eons it took to create the fossil fuels. Even those who doubt the science behind global warming must have some sense of worry—perhaps this natural system called the atmosphere cannot cope with such a precipitous change.

CAN THINGS GET WORSE?

Recent scientific studies from the Greenland Ice Core Project suggest that matters may get much worse, and much faster than previously thought possible (Lynch, 2002; Bryson, 2003). The project carefully dated each of the ice layers, and the scientists found that at the end of the last ice age—about 12,000 years ago—there were very sudden shifts in temperature, with swings between hot and cold by as much as 8 C° (14.4 F°) in periods of about ten years. The cause is thought to be meltwater from Greenland flowing in front of and shutting down the Gulf Stream in the north Atlantic. This is part of a worldwide system of ocean currents known as the "thermohaline conveyer belt", which is driven by differences in temperature and salinity (that is, density). These temperature variations were likely restricted to the North Atlantic region, and were not global. It is possible that global warming today, with its melting of the Arctic and Greenland ice sheets, may trigger a similar response. At the end of the previous ice age, humans were migratory and few, so they could survive and adapt. But today such rapid and sweeping thermal changes would have dire consequences.

In addition to finding that the global average temperature for the latter part of the twentieth century is at least as high as at any time in the past 600 years, the IPCC also found that the temperature trend has been upward since 1900. Furthermore, the geographic patterns of the global temperature trend are unlike patterns associated with natural climate variation. Instead they are similar to what is predicted by global warming, and the magnitude of changes observed is very similar to that estimated by computer modeling (Gelbspan, 1997).

Studies at numerous organizations throughout the world support the IPCC findings. In 1995, scientists at the US National Climate Data Center found a pattern suggesting the US climate had become warmer. They calculated a "greenhouse climate response index" based on factors such as higher than normal daily minimum temperatures, extreme or severe droughts in warm months, and much higher than average precipitation in cool months. The index has been consistently higher since the late 1970s. In 1995, a research team from Germany's top climate laboratory, the Max Planck Institute for Meteorology, concluded that there was only 1 chance in 40 that natural variability could explain the warming of the past three decades (Leggett, 1999).

There are many other signs that global warming has begun in earnest. In addition to historical meteorological records, indirect evidence has been accumulated from ocean temperatures, melting mountain glaciers, rising sea levels, shifts in distribution of some species of animals and plants, tree rings, ocean corals, and underground temperatures. All of these show substantial warming over the past century.

Scientists from the University of Texas performed a massive compilation of natural history data. Their results, published in 2003 in the journal *Science*, showed no clear trends before 1950, but in the years following they found strong patterns in species' distributions, including a poleward movement of about 6 km (3.7 mi.) per decade, a movement up mountainsides by 6.1 m (20 ft) per decade, and an advance of spring by 2.3 days per decade. Although these amounts may seem small, the rates are—in the natural world—very rapid. Their impacts on wildlife and plants can be devastating. So strongly did the results correlate with global temperature changes that they have been hailed as constituting a globally coherent "fingerprint of climate change" (Flannery, 2005).

Records indicate that the twentieth century was the warmest century of the millennium. The 1980s brought North America the five hottest summers of the previous century (Mitchell, 1991). But the heat wave continued. The year 1995 turned out to be the hottest year since records began. The increased temperatures were accompanied by severe weather. Insurance costs for catastrophic property loss also rose to record highs during this period. Then 1998 broke all records for global average annual surface temperature by a frightening margin (Leggett, 1999).

And the heat goes on and on. In 2007, the IPCC reported that 11 of the 12 hottest years since modern record-keeping began in the 1850s occurred from 1995 to 2006; in descending order, the years with the highest global average annual temperatures were 2005, 1998, 2002, 2003, and 2004.

CORAL BLEACHING

Of all the treasures found in the oceans, none is more diverse and alive with life and stunning colorful formations than a coral reef. Acting as nurseries for a wide variety of fish and as major tourist attractions, coral reefs around the world provide about $30 billion in income each year. Yet these beautiful and bountiful formations are on the brink of collapse.

Corals are very temperature-sensitive organisms, thriving in waters of about 27°C–28°C (81°F–82°F), but suffering when the temperature rises even a degree above this level. Heat-stressed corals expel the algae that live within their bodies and which provide their food supply. They turn white—hence the term "coral bleaching". If stressed for too long, the corals die. During the 1980s and 1990s, every major coral reef system in the world was affected by bleaching at one time or another. Although other factors can cause bleaching, it was shown that many of the recent episodes were associated with waters that were warmer than usual (Leggett, 1999). Ocean temperatures have continued to rise since then.

The IPCC conclusions have been endorsed by more than 100 Nobel laureates, by 17 national scientific academies, and by most of the world's governments (including the United States). In 1997, for example, 1,500 scientists—along with 104 of the 138 living winners of Nobel prizes in the sciences—signed a declaration urging world leaders to cut greenhouse gas emissions immediately (Leggett, 1999).

The question is not if global warming will happen; the question is, when and by how much? Skeptics may argue about the fine details, but the overall conclusion is sound.

The Consequences

As we will see, the consequences of greenhouse gases are predicted to be profound. This may seem surprising since these gases constitute only about 0.04 percent of the atmosphere. The global climate system has evolved over a long time to be incredibly complex, with its many components working together in a closely interconnected and symbiotic equilibrium. If one component, no matter how small, is suddenly altered, it will affect the overall system. It is the same with the human body; changes in such small entities as enzymes, white blood cells, or trace chemicals can dramatically affect an individual's health and well being.

Climate

The most alarming consequence of global warming is that it may cause instabilities in the climate system, upsetting the balance between winds, clouds, rainfall, surface temperatures, ocean currents, ice caps, and sea level. The result will be increasing extremes of temperatures and storms. As the IPCC report stated in technical language, "When rapidly forced, nonlinear systems [like the climate] are especially subject to unexpected behavior." Changes in climate will, of course, have a direct impact on vegetation and animal life on both land and in the sea.

Consider precipitation. A rise in global temperature will cause a widespread increase in the amount of water moving through the evaporation-precipitation cycle. This would happen because a higher temperature increases evaporation and transpiration and also raises the capacity of the atmosphere to hold moisture. At the same time, higher temperatures cause the atmosphere to circulate more actively. This will increase the frequency of very heavy and extreme rainfalls and of local flooding.

The number of thunderstorms and their intensity will increase, leading to more lightning. Research at the Goddard Institute for Space Studies showed that a 1 C° (1.8 F°) rise in global temperature results in a 6 percent increase in lightning (Francis and Hengeveld, 1998). This, in turn, will cause more forest fires, especially if anticipated summer dry spells become more prevalent. In addition, thunderstorms breed hail and tornadoes. It has been shown that the monthly frequency of tornadoes in the Canadian prairie provinces correlates closely with the average monthly temperature, suggesting that global warming will bring increased tornado activity to this region. Hurricanes, which form and grow over ocean water with a temperature of 27°C (81°F) or greater, are predicted to become more common. At the other extreme, many regions will experience more frequent, prolonged, and severe droughts due to more rapid evaporation.

Scientists at the National Climate Center of the US National Oceanic and Atmospheric Administration (NOAA) found that extreme weather events including droughts and floods in the United States, Russia, and China have been increasing (Gelbspan, 1997). They developed a "climate extremes index" for the contiguous United States, which incorporates temperatures, precipitation, drought severity, and the wind velocities of landfalling tropical storms and hurricanes. The index has increased steadily and significantly from 1970 to 2007 (NCDC, 2008).

As the global average temperature rises, it is logical to expect the number of extremely hot days to increase. Mathematical analyses have shown, however, that the increase would be out of all proportion to the increase in temperature. A global rise of 1.6 C° (2.9 F°) would change the frequency of an extremely hot summer in England from about once in 75 years to once every three years—a factor of 25! A climate model showed that an increase of 4 C° (7.2 F°) in Toronto's average temperature would increase the probability of summer days with temperatures exceeding 30.5°C (87°F) from one in ten to about one in two (Francis and Hengeveld, 1998). Studies for other localities have produced similar results.

A quantitative method of measuring the effect of climate change is through insurance claims for weather-related disasters. Such claims jumped from about $5 billion in the 1980s to about $30 billion in just the first half of the 1990s (Gelbspan, 1997). From 1965 to 2003 natural catastrophes, most of them weather related (tornadoes, heat waves, forest fires, and severe floods) caused losses that soared from about $1 billion per year to $60 billion per year. Although these losses are also related to population and economic growth, an increase in extreme events is playing a significant role. Insurance firms are taking global warming very seriously. In fact, Swiss Re, the world's second largest reinsurer, announced in 2003 it is considering denying coverage to those companies who are not doing enough to reduce their emissions of greenhouse gases (Flannery, 2005).

National defense, insurance policies, and many other human activities design their strategies to account for worst-case situations. In a similar fashion, society should take out "insurance" against extreme future climate changes. This is important because scientists have identified a number of processes triggered by warming that cause additional warming, thus accelerating the entire process. This could result in much bigger temperature rises than the ones currently predicted. In one projection, known as the "Methane Doomsday Bomb" scenario, warming of the Arctic climate would release methane hydrates (ice-like substances that form under pressure in permafrost and sub-sea sediments), so that a massive amount of methane—a powerful greenhouse gas—enters the atmosphere. In another scenario, higher temperatures might cause large areas of forests (an immense carbon reservoir) to die, resulting in further carbon dioxide releases and further warming (Leggett, 1999).

Rising Sea Level

Because global warming is felt most at the polar regions, there would be widespread melting of the polar ice caps with an attendant rise in sea level and flooding. Thermal expansion of the oceans also contributes to a rising sea level; water expands as it is heated. The IPCC in 2007 reported there would be an approximately 18 to 59 cm (7.1 to 23 in.) rise in sea levels by 2100 depending on future population and economy scenarios. Although these projections may seem small they would have dire consequences on the Netherlands, Bangladesh, low-lying islands in the Pacific, and elsewhere.

In a worst-case scenario, a 1 m (3.3 ft) rise in sea level would drown 25–80 percent of present US coastal wetlands, mostly in the southeast in Florida, Louisiana, and North Carolina. The Gulf of Mexico would surge into Louisiana as far as 45 km (30 mi.) inland (Mitchell, 1991).

A NASA scientist reported that the Arctic ice cover is melting faster than expected and will all be gone by 2100. Between 1978 and 2000 approximately 1.2 million km^2 (0.5 million sq. mi.) of permanent ice melted. That represents a loss of about 9 percent of the ice cover per decade (Comiso, 2002). The melting correlates with the increase of about five days per decade in the length of the summer melting season (Smith, 1998). As the ice melts, global warming increases—because less solar heat is reflected off the snow's white surface.

Agriculture

Some have argued that global warming would benefit agriculture because an increase in carbon dioxide stimulates photosynthesis, thus improving plant growth. This is known as the "carbon dioxide fertilization effect". Also, a warmer climate helps plants grow faster and larger. It has been shown that a doubling of carbon dioxide can, under ideal circumstances, increase wheat and rice yields by up to 25 percent, and soybean yields by up to 49 percent (Melillo et al., 1993).

Overall, however, the effects on global agriculture would be negative. With a warming climate, many crops would need to migrate to new regions, a difficult transition that would require considerable time and would be severely restricted as a result of existing land use. Researchers at the University of Oxford performed modeling studies that showed most of the world's important food crops would be affected by global warming. They estimated that there would be a decrease in grain yields of 10–15 percent in South America, Africa, and southeast Asia (Leggett, 1999). Rising temperatures, water shortages, and increased droughts could have devastating effects in major food-growing areas such as prairie farmlands.

Diseases

Global warming is also expected to increase the transmission of infectious diseases such as malaria, dengue fever, and yellow fever by expanding the ranges of carriers of these diseases. The rapid advance of West Nile disease, carried by mosquitoes across North America, may just be a harbinger of things to come. With warmer temperatures, not only does the range of mosquitoes increase, but their breeding cycle also accelerates—in other words, there are far more of them. One study estimated that a temperature increase of several degrees by 2100 could multiply the ability of mosquitoes to transmit diseases 100 times in temperate climates (Dotto, 2003).

Is Doomsday Nearing?

In the past when ice ages rolled their glaciers over the land, humans and other plants and animals were able to migrate due to the slow progression of the changes. But the ability to migrate in the future will be greatly restricted; this is not only a result of the rapidity of climate change, but also because almost all habitable places on the overpopulated Earth are already occupied. In addition to major disruption to food supplies, there would be enormous strife and, quite possibly, war. As described by Jared Diamond in his book *Collapse* (2005), the 1994 bloodbath in Rwanda was largely caused by a very high population density and an insufficient food supply.

By establishing the IPCC in 1988, the United Nations and the international community formally acknowledged climate change was a serious problem and required global action. But in spite of this positive start, meaningful action to resolve the global warming problem has been much more difficult—if not impossible—to achieve.

The Energy Dilemma

A major reason for the difficulty in taking action against global warming is that energy, its major cause, provides the basis for a high standard of living. In North America, our standard of living is built around energy; we are addicted to it, mainlining on it. That is the crux of the problem. This energy is provided largely by burning fossil fuels, which has released and continues to release enormous amounts of carbon dioxide into the atmosphere where it slowly traps heat and warms the globe. Since 1950, the use of fossil fuels has increased over 30 times.

Energy is also central to developing countries in the struggle to raise their standards of living. China, for example, has the ambition for each household to possess a refrigerator. Even if these refrigerators are the newest, most energy-efficient models, they would require an additional electrical capacity of 20,000 MW, most of which would be supplied by coal-fired power stations. And this does not take into account automobiles, stoves, televisions, other appliances, or the power needed by the factories that manufacture these goods.

Projections indicate that by 2040 the world will need three to four times the electricity capacity that we had in 2000—an alarming increase in energy demand.

IT'S ALL ABOUT ENERGY

The numbers involved with global energy are so large that they are often difficult to comprehend. To help put them into perspective, consider the mammoth Three Gorges hydropower project in China. When completed in 2009 it will be the world's largest hydro installation consisting of 26 separate 700 MW generators for a total capacity of 18,200

MW. This is an enormous energy source that exceeds, for example, the entire capacity of Arkansas (14,500 MW) and almost matches the capacity of the province of Ontario, which in 2006 was about 26,740 MW (supplied by 13 nuclear reactors, five coal-burning stations, 64 hydro facilities, and three wind-power stations). China's projected growth in electricity demand of 4.3 percent per year from 2003 to 2025 is such that the Three Gorges project will satisfy less than two years' growth. In other words, China will need to build the equivalent of a new Three Gorges project—about 18 large coal and nuclear plants—every two years! With abundant coal reserves, most of China's demand will be met by coal-fired power stations—lots of them.

The Rich-Poor Hurdle

A solution to global warming is complicated by the vast gulf between rich and poor nations. In 2004, the annual emission of carbon dioxide per capita was about 4.5 tonnes (5.0 tons). There was, however, a major difference between developed countries, whose average was about 25 tonnes (28 tons), and developing countries, whose per capita average was about 3 tonnes (3.3 tons). In other words, the developed countries, with about 20 percent of the world's population, generate about 80 percent of greenhouse gases. And no country is more of an energy hog than the United States. Canada is also high on the list in per capita carbon emissions. In 2004, the United States and Canada together accounted for roughly a quarter of global emissions of carbon dioxide—although they have only about 5 percent of the world's population.

China insists that it will accept no mandatory limits on its carbon dioxide emissions; limits would almost certainly reduce its industrial growth. It argues that rich countries caused global warming and should find a way to solve it without impinging on China's development (Kahn and Yardley, 2007). Even if industrialized countries manage to reduce their carbon dioxide emissions by 50 percent, nothing will be achieved. This is because the population and economic growth in developing countries will drive up annual emissions to more than double today's rates (Mitchell, 1991).

Yet we cannot condemn developing countries to eternal poverty; ethically, we are bound to help them improve their quality of life, which will increase their energy demand and, therefore, their greenhouse gas emissions.

The wealthy nations must not only make their own cuts, they must reduce their consumption rate to compensate for poor nations. Are we willing to do this? To date, the world's leader, the United States, has not shown the slightest interest in making such sacrifices.

The Kyoto Protocol

In spite of these difficulties, a major step was taken when the United Nations Framework Convention, known as the Climate Convention, was signed by more than 160 nations at the Rio de Janeiro Earth Summit in 1992. It came into force in 1994. The main objective was for developed countries to take the necessary actions to reduce greenhouse gas emissions, especially carbon dioxide, back to 1990 levels by 2000. The long-term goal was to stabilize concentrations of greenhouse gases at levels that would prevent dangerous interference with the climate system.

In spite of the promises, world greenhouse gas emissions increased 26 percent during this period. This disappointing development shows that prospects for the future are bleak indeed. The enormous juggernaut of population and its thirst for energy continues to gather momentum, and the means to slow it—much less to reverse it—are proving difficult to find.

As expected, the strongest support for action was taken by those that had the most to lose, a group of island nations—the Alliance of Small Island States, consisting of more than 30 countries from the Pacific and the Caribbean including the Philippines, Jamaica, the Marshall Islands, and Samoa (Gelbspan, 1997; Leggett, 1999). As UN Ambassador Tuiloma Slade of Samoa stated, "The strongest human instinct is ... survival ... and we will not allow some to barter our homelands, our people, and our culture for short-term economic interest. It is absolutely necessary to act." He was referring to the projected sea level rise that would devastate these low-lying nations.

The next major step was the Kyoto Protocol, which was concluded in 1997 and committed developed countries to reduce their greenhouse gas emissions to 5 percent below 1990 levels by 2008–2012. No reductions, however, are required by developing countries. This creates a major problem because the greenhouse gas emissions of China and India are growing so rapidly.

A significant deficiency is that Kyoto targets are much diluted from what experts believe is actually necessary. At a preliminary meeting in 1997 in Bonn, nearly 150 countries asked for reductions of 15 percent below 1990 levels by 2010, with 35 percent reductions by 2020. Some experts believe that rich nations need to make cuts of 90 percent by 2030 if we are to avoid a 2 C° (3.6 F°) rise, a threshold that could trigger the collapse of major ecosystems (Monbiot, 2007). But intense lobbying, obstruction, and filibustering by the oil-rich nations and the United States resulted in a much weakened version of what had been originally proposed. Now the international community cannot meet even the water-downed targets.

Numerous lobby groups have played a vigorous role in opposing any meaningful action in resolving global warming. These lobbyists represent influential organizations that have a vested interest in fossil fuels such as the auto, steel, oil, and petrochemical industries; electric utilities; and a host of others. In the US system of government, such lobby groups have inordinate influence. The strategy of these groups, as seen with acid rain and the ozone hole, is first to deny that the phenomenon exists at all. Then there are calls for study, more study, and ever more study, to resolve extended disputes over minor scientific uncertainties. Finally, there are predictions of economic collapse and doom should corrective measures be implemented.

To encourage carbon emission reduction, "emissions trading rights" were introduced. In this scheme, carbon credits would be allocated among the various countries and then traded as needed. For example, if the emissions of one country exceed their target they can purchase credits from another country that has reduced their emissions below their target. In this way, market forces come into play, and there is strong economic motivation to reduce emissions as far as possible, not just to the regulatory limit. The carbon trading system is similar to that used successfully for sulfur dioxide (see Chapter 6).

The Chicago Climate Exchange was created in 2003 to trade carbon credits in North America, and although it operates on a voluntary basis since the US has not joined Kyoto, many large companies have joined, including the Ford Motor Company, International Business Machines, and Abitibi Consolidated. In 2005, total value of credits traded in the Chicago Exchange was $2.8 million, considerably less than the $8.2 billion traded on European and other exchanges, where Kyoto targets are enforced by the European Union (Marotte, 2006).

TABLE 9-1: Carbon Dioxide Emissions from Fossil Fuels by Country (EIA, 2005)

Country	CO_2 (million tonnes)		% increase	% of world total, 2004	CO_2 (tonnes) per person for 2005
	1990	2005			
United States	5,013	5,957	18.8	21.1	20.1
China+	2,241	5,323	110.0	18.9	4.1
Russia	n.a.*	1,696	n.a.	6.0	11.8
Japan	1,015	1,230	21.2	4.4	9.7
India+	588	1,166	98.3	4.1	1.1
Germany	980**	844	-13.9	3.0	10.2
Canada	479	631	31.7	2.2	19.2
United Kingdom	598	577	-3.5	2.0	9.6
Iran	202	451	233	1.6	6.6
Australia	263	407	54.8	1.4	20.2
Saudi Arabia	208	412	98.1	1.5	15.6
Brazil	223	361	61.9	1.3	1.9
Poland	331	285	-13.9	1.0	7.4
Denmark	57	51	-10.5	0.2	9.4
Zimbabwe+	15	12	-20	0.0	0.9
World Total	21,426	28,193	31.6		

* Russia was still the USSR in 1990

** East and West Germany combined

+ Non-developed countries, not subject to Kyoto Protocol

MAKING THE KYOTO PROTOCOL

The international meeting to hammer out a global solution to the climate change crisis was no small affair. It took place at a large conference center in the north end of the ancient city of Kyoto, set among hills and lakes. The peaceful and beautiful setting formed a marked contrast to the bitter lobbying, negotiating, and backroom politicking that went on inside the center. The media seemed to realize that the Earth's future hung in the balance. Approximately 1,500 government delegates were in attendance;

they were swamped by about 5,500 journalists from 400 media organizations. Over 3,600 observers including intergovernmental agencies and non-governmental environmental and business organizations were constantly lobbying and advising. The Greenpeace delegation alone was 45 strong. The halls were too small to seat all attendees, so much of the action was followed on closed-circuit television (Leggett, 1999).

To become international law, the protocol had to be signed by at least 55 developed countries whose cumulative greenhouse gas emissions in 1990 formed at least 55 percent of the world's total. Canada signed the Kyoto agreement in late 2002, to become the 99th nation to join. But without the ratification of the United States, Russia's signature was necessary. In 2004, Russia reversed its previous position and, much to the delight of European countries and environmentalists, signed the Kyoto Protocol, bringing it into force on February 16, 2005. Initially, Australia refused to sign the Kyoto Protocol. In December 2007, Australia's newly elected prime minister Kevin Rudd reversed the previous policy and committed his country to joining the agreement.

Table 9-1 provides data on carbon dioxide emissions from fossil fuels by different countries. (Note that the carbon dioxide emissions listed here represent only about 60 percent of total greenhouse gases. This is because methane, halocarbons, and nitrous oxide are not included.) It illustrates several important points. First, it shows that two countries, the United States and China, are by far the largest carbon dioxide emitters, accounting together for 40 percent of the world's emissions. In 2007, the Netherlands Environment Assessment Agency announced that China had surpassed the United States in greenhouse gas emissions (Kahn and Yardley, 2007). The highest energy users on a per capita basis are the United States, Australia, and Canada.

This figure also shows disappointing results: instead of decreasing, almost all countries have increased their emissions significantly since 1990, and the total world carbon dioxide emissions were 31.6 percent higher in 2005 than in 1990. With coal mining, energy demand, and populations increasing, this trend will not change in the near future.

Only some European countries, notably Germany, Russia, Poland, and the United Kingdom, have decreased their emissions. These countries fall into two camps. The first group comprises Russia and the eastern European countries like Poland and East Germany, which suffered a severe economic downturn with the end of the Cold War and the fall of the iron curtain in 1989. This resulted in reduced energy consumption and a consequent decrease in carbon dioxide emissions. As these countries regain economic vitality, their emissions will increase.

The second group includes western European countries such as the Scandinavian nations, Italy, the Netherlands, Belgium, and the United Kingdom, who have made the real progress in slowing carbon dioxide emissions. Here, two important factors have contributed: first, population growth in these countries has slowed significantly or stopped, and second, there is a much greater awareness and concern for the environment. For example, Denmark—probably the most proactive country—quickly enacted a carbon tax and enforced efficiency standards for its utilities with the goal of attaining a 20 percent emission reduction below 1990 levels by 2005, far exceeding the Kyoto requirement (Gelbspan, 1997).

Canada failed to decrease its emissions. In fact, as a result of vigorous economic growth, population increase, increasing exports of energy to the United States, and rapid development of the tar sands, Canada actually increased them by 32 percent between 1990 and 2005. In 2006, the Canadian government admitted that it could not meet its Kyoto commitments and was investigating ways of withdrawing from the agreement. At the United Nations climate conference at Nairobi in November 2006, the Canadian position was ridiculed. A European-based development group placed Canada in 51st position of 56 countries ranked for their climate-change policies. The best scores were given to Sweden, the United Kingdom, and Denmark. Joining Canada at the bottom of the heap were Australia (48th), the United States (53rd), and China (54th).

As the largest greenhouse gas emitter and also the richest and most technologically advanced country in the world, the United States should have taken a leadership role at Kyoto. Instead, it was one of the biggest obstructionists. In the spring of 2001, the United States announced that implementing the Kyoto Protocol would be harmful to its economy and it would not sign. Instead, the United States announced an initiative to reduce "greenhouse intensity" by 18 percent between 2002 and 2112. "Greenhouse intensity" is defined as the amount of carbon dioxide emitted per dollar of GDP. Since the United States is committed to increasing its GDP, this initiative states that it will continue increasing its carbon dioxide emissions, only at a lower rate. The option of making an absolute emission reduction or returning to 1990 levels is not on the table.

The withdrawal of the world's most powerful country from the protocol places pressure on the committed countries. They are in danger of losing their competitive edge on the world marketplace if the cost of their energy supplies—and hence the cost of their commodities—are driven higher because of meeting Kyoto responsibilities.

Since its withdrawal the United States has embarked on an aggressive program to build further coal- and natural-gas-fired power plants. This is in spite of a *New York Times* public opinion poll that found fully 65 percent of Americans wanted the United States to take immediate steps to curb its emissions, regardless of what other countries did. As the *Times* stated, "The American people are far more willing than their government to take early, unilateral steps."

There are, however, some glimmers of action at the state level. In late 1993, seven northeastern states (Connecticut, Delaware, Maine, New Hampshire, New Jersey, New York, and Vermont) formed the Regional Greenhouse Gas Initiative to reduce carbon dioxide emissions. Maryland joined in 2007; the District of Columbia, Massachusetts, Pennsylvania, Rhode Island, and the eastern Canadian provinces are observers in the process. Central to their initiative is the implementation of a multi-state cap-and-trade program with a market-based emissions trading system. The proposed program will require electric power generators to reduce carbon dioxide emissions.

In 2006, California's legislature approved the broadest restrictions on carbon dioxide emissions in the nation. The California bill requires a 25 percent cut in carbon dioxide pollution produced within the state's borders by 2020 in order to bring the total down to 1990 levels. The legislation also provides a statewide market system designed to make it easier for heavily polluting industries to meet the new limits. They would be able to buy credits from companies that emit lower emissions than the caps allow.

But it is difficult to see how the Kyoto Protocol will solve global warming for a number of reasons: the Kyoto targets are much lower than required (reductions of 70 percent—not 5 percent—are needed by 2050 to keep carbon dioxide at twice the pre-industrial level); the United States has refused to sign the Kyoto Protocol; developing nations are exempt from making emission reductions; most developed countries have increased emissions since 1990; and national economies, with their attendant energy consumption, are projected to grow rapidly in the future.

Solutions: Applying Technology

Global warming is one of the most challenging problem ever faced by humanity. As though raging storms and rising seas weren't enough, there are many additional reasons to combat global warming. We have seen that coal-fired power plants emit a plethora of pollutants in addition to carbon dioxide. Reducing greenhouse gas emissions will bring substantial benefits to public health by reducing local and regional air pollution; global warming is inextricably linked to deadly smog and acid rain.

But reducing greenhouse gases will be extremely difficult. We cannot rely on any one technical solution; rather, the battle must be fought relentlessly on many fronts. Let us see what armaments are at our disposal. My purpose is not to provide a quantitative analysis of different methods and their potential for reducing greenhouse gases. Instead, I will briefly describe the methods that are available.

The methods for decreasing greenhouse gas emissions are many, and they fall into two broad categories: technological and social. The next section describes the former while the latter are discussed in Chapter 11.

Energy Efficiency and Conservation

The simplest way to reduce greenhouse gases is to use less energy. As a society and as individuals, we need to respect energy and do all we can to conserve it. This includes improving insulation in buildings, turning down thermostats in winter and turning them up in summer, carpooling, making vehicles and appliances more energy-efficient, and using mass transit.

We have been down this road before. In response to the energy crisis of the 1970s, when OPEC dramatically raised the price of oil, a wide array of energy conservation measures was successfully introduced. New fuel-efficient engines were developed, people drove more slowly, and thermostats were lowered. Unfortunately, in North America we have slipped back into old habits. Europe, however, retained its conservation ethic, due to a greater concern about dependence on foreign oil supplies. The result is that western European countries consume about 50 percent less energy per person than in North America, and they do this while enjoying a high standard of living. North America must copy the European ethic.

Recycling saves energy and must be promoted vigorously. Recycling 20 aluminum cans, for example, saves the energy equivalent of 10 liters (2.6 gallons) of gasoline. Studies and experience, however, have shown that even the best recycling programs can only achieve about 50–60 percent diversion of municipal waste from disposal. Instead of the remainder going to landfills, it should be incinerated to produce electricity. In other words, nonrecyclable garbage should be considered a renewable resource to replace the burning of fossil fuels. Incineration of waste to create energy is widely practiced in Europe, less so in the United States, and hardly at all in Canada.

One energy source, albeit a small one, that is currently being mostly wasted is the methane seeping from thousands of landfills and sewage treatment plants. The technology for capturing and burning this gas to generate electricity or heat buildings is available and has been successfully applied. Of the approximately 13,000 active and closed municipal landfills in the United States, only 445 were capturing methane for conversion to energy in 2007 (EPA, 2008c).

Smart growth, which encourages compact, complete communities with mass transit (see Chapter 5), can play an important role in reducing carbon dioxide emissions from transportation and in decreasing our reliance on the automobile.

A broad range of incentives needs to be established by governments to promote conservation. In particular, the price of gasoline, which is cheap in North America—even after the recent large price spikes—should be increased. Other measures could include enforcing better gasoline efficiency in new vehicles, taxing gas-guzzling vehicles such as SUVs, and so on. Two major steps would be as follows. First, remove subsidies from fossil fuels, which are estimated to cost US taxpayers $15 to $35 billion per year for oil and gas alone (Kachan, 2007). Second, apply a carbon tax to fossil fuels. Such monetary penalties are very effective in reducing consumption. Although some argue that carbon taxes can be difficult to manage and are more punitive to some industries than others, they have been successfully implemented in Denmark, the Netherlands, Sweden, Finland, and Norway (Willis, 2006). The United States and Canada have so far rejected the use of carbon taxes, although British Columbia introduced carbon taxes in 2008.

Regulatory-imposed guidelines do work. In 1987, for example, the US Congress passed an Appliance Energy Conservation Act requiring manufacturers of refrigerators, freezers, water heaters, washers and dryers, dishwashers, and air conditioners to meet specified targets for energy efficiency. It is estimated that this act saved about $20 billion in electricity by 2000 (Young, 1990), with an attendant decrease in emissions.

Converting to Renewable Energy

Renewable energy sources—such as windmills, solar panels, and hydroelectric generators, which produce no greenhouse gases—are an excellent way of combating global warming. The use of wind power, a seemingly simple and benign energy source, has increased rapidly in recent years as wind turbine technology. From 2000 to 2007, worldwide wind-power generating capacity increased from 18,000 MW to 92,000 MW, a five-fold increase. The world leader in wind power is Germany, followed by the United States, Spain, India, and Denmark. Denmark produces almost half of its electricity from wind (Brown, 2008).

Wind power has practical limitations, however, because it is intermittent and diffuse. To replace a large fossil fuel burning plant would require thousands of turbines, which would be costly and require a large land area. The NIMBY ("Not In My Back Yard") syndrome would be a major obstacle to the siting of such enormous wind farms. An infrastructure to collect and store energy for periods when the wind is not blowing is also required. Finally, wind farms of this magnitude need many lines to collect and transport the electricity to a few central points where the energy can be supplied to high-voltage transmission lines for sending to cities.

Ultimately, solar power must play a significant role. Although solar energy can be used to heat homes and water directly, the most flexible use is with "photovoltaic cells", which turn sunlight directly into electricity. The photoelectric effect was discovered in 1839 by a French scientist, Edmond Becquerel. He found that when he exposed two different brass plates immersed in a liquid to sunlight, it produced a continuous electric current. In the late 1870s, English scientists found they could obtain the same result using selenium. It was not until 1954, however, that the first practical photovoltaic cell using silicon was developed at Bell Laboratories (Komp, 1995). Photovoltaic cells have many benefits: they are rugged, reliable, have no moving parts, can withstand extreme temperatures, are impervious to most corrosive chemicals, and give off no toxic emissions. Solar cells were first used, and continue to be extensively used, in the space program for satellites where there is continuous sunshine.

The first silicon cells converted about 4 percent of the incoming solar energy into electricity. Although efficiency has improved significantly and today cells convert about 16 percent, there are still major hurdles to overcome before photovoltaic cells can compete with the lower cost of electricity generated from fossil fuels. The methods for growing silicon crystals have evolved enormously and individual crystals 15 cm (5.9 in.) in diameter and 2 m (6.6 ft) long are routinely grown (Komp, 1995). Individual crystal wafers are about 300 microns thick. Considerable research is underway on developing different crystal systems, lowering production costs, finding ways of gathering and focusing more sunlight on the cells, and other technical innovations.

The use of photovoltaic cells is gaining increasing acceptance, particularly in applications where there is no access to electrical grid lines. Typical uses include powering lights on buoys and lighthouses, providing electricity for sailboats, pumping water from aquifers for cattle, lighting temporary signs along highways, and powering electric-vehicle charging stations. More government funding and regulation are required to help bring solar cells to market. A large step forward, for example, would be building codes that require all new buildings to have solar panels. This would very quickly create a large market that would drive down costs and improve quality and service.

Because the intensity of the sun decreases on cloudy days and disappears at night, the use of solar energy for large-scale power generation is complicated by the need for storing energy. Furthermore, large-scale solar power generation would require immense fields of solar panels, again triggering NIMBY opposition.

Cities, with their densely packed apartments, factories, office towers, theaters, shopping malls, and subways, consume enormous quantities of energy. Large power plants are needed to supply this demand. Renewable energy sources such as solar and wind power, although excellent sources of clean energy (particularly in rural and residential settings) will not be able to make more than a minor contribution in high-density urban settings.

Hydroelectricity has been a major source of energy for over a century. In Canada it provides about 11 percent of total primary energy.

Worldwide, about 60 percent of hydro potential has been exploited, and in industrialized countries the exploitation is close to 100 percent. There is little potential to replace fossil-fuel power by increasing hydropower capability. In those places where hydro potential does exist, the flooding of reservoirs for hydropower developments is becoming increasingly difficult due to conflicts over land use. Even small contributions from hydro, however, should be pursued. For example, run-of-the-river turbines that do not use dams can be installed without the need for environmentally damaging reservoirs.

Renewable energy sources must be pursued vigorously. But we should not believe what many environmentalists naively trumpet: that conservation and renewable energy alone will solve the global warming problem. The global energy appetite is voracious and consumption is projected to increase by an enormous 58 percent from 2001 to 2025 (EIA, 2003). The problem is two-fold: replacing existing energy infrastructure with clean power, and creating new infrastructure to meet future needs. Even with vigorous expansion, it is virtually impossible for renewable energy to make more than a small, however important, contribution. Much more is needed.

Clean Fossil Fuels

There is enormous pressure to develop clean fossil-fuel energy. This is an appealing goal, for it would avoid the difficulty of changing to a renewable-energy economy, and would allow "business as usual". Reducing emissions from coal, in particular, is the focus of considerable research because coal reserves are so widespread. Let us see how fossil fuels, especially coal, can be cleaned up.

Conversion to Natural Gas

The simplest approach is to switch from coal to a less polluting fossil fuel. Generating one kilowatt-hour of electricity by natural gas produces about 50 percent less carbon dioxide than by coal, and about 45 percent less than by oil. For this reason, and also to reduce the emission of sulfur dioxide and other noxious gases, many utilities and industries around the world are switching to natural gas. In addition, most new power stations are designed to use the latest natural gas combined-cycle gas turbines (see Chapter 7).

This approach is effective. In 1998, two international carbon giants, BP and Shell, committed to cutting their greenhouse gas emissions 10 percent below 1990 levels—cuts deeper than what the Kyoto Protocol requires (Leggett, 1999). Also, by 2002, the Canadian Steel Producers Association, which represents all 17 steel mills in the country, had reduced their collective emissions by 16 percent. These reductions were made by converting from coal to natural gas and by instituting conservation measures (CSPA, 2001).

Although a positive step, the use of natural gas is only a temporary solution, for reserves of this resource are dwindling and its escalating price will soon make this option economically unattractive.

Carbon Dioxide Capture and Sequestration

At present no reliable and cost-effective methods are available for removing carbon dioxide from the emissions of fossil-fuel plants. Research is underway to find better methods.

One method of removing carbon dioxide from emissions is by "scrubbing", that is, spraying power-station exhaust gases with solutions containing solvents called "amines", preferably monoethanolamine. (Amines are used to collect the carbon dioxide that makes soft drinks fizzy.) Amines are not without problems, however; they degrade in oxidizing conditions and require considerable energy for regeneration (Audus, 2000).

In the early 1980s carbon dioxide from a coal-fired power station in western Canada was captured and used for enhanced oil recovery. Carbon dioxide was injected into oil reservoirs from which oil could no longer be withdrawn by conventional methods. The carbon dioxide "pushed" the remaining oil toward wells where it could be pumped to the surface. The carbon dioxide extraction proved to be prohibitively expensive and solvent degradation posed a significant operating problem, so the project was abandoned.

A large pilot project program is currently being conducted by the University of Regina to improve this carbon dioxide extraction method and make it cost effective. Various solvents are being tested at a nearby coal-fired power station.

Once captured, carbon dioxide must be safely and permanently disposed of—a process called "sequestration", which has its own complexities. The main methods under consideration involve injecting carbon dioxide deep underground into formations such as saline aquifers, which have no value for drinking water or irrigation.

There are major uncertainties regarding the deep storage of massive quantities of gaseous carbon dioxide. The short-term risk is that the carbon dioxide might escape to the surface quickly where, being denser than air, it can flow downhill and collect in pockets, asphyxiating people and animals and damaging plants. In 1986, for example, a large quantity of carbon dioxide was suddenly released from the waters of Lake Nyos in Cameroon, where it had collected from underlying volcanic vents. The cloud of carbon dioxide traveled over 10 km (6 mi.) and killed 1,700 people and thousands of cattle (Keith, 2002). At Mammoth Mountain ski resort in California, skiers have died when they entered low hollows where carbon dioxide, which seeps naturally from the ground in that area, had accumulated (Goodell, 2006).

Even if short-term leakage and environmental damage can be avoided, the carbon dioxide might escape slowly over years or decades, re-entering the atmosphere and negating the efforts to reduce global warming.

The processing of natural gas has provided valuable information on sequestration. Raw natural gas contains several impurities but primarily carbon dioxide (about 2.5 percent) and hydrogen sulfide—"acid gas" (H_2S). In the past, the carbon dioxide and acid gas were flared into the atmosphere, but environmental concerns have stopped this practice and led to the recovery and sale of the sulfur. But the carbon dioxide continues to be released into the atmosphere. In recent years the market for sulfur has weakened and many producers are injecting the acid gas and carbon dioxide deep into suitable geological formations (Keith, 2002). Not only have relatively large amounts of carbon dioxide been prevented from entering the atmosphere, but industry has obtained valuable experience with this sequestration method. Results suggest it could be implemented on a much larger scale.

Large quantities of hydrogen are used by the oil and gas sector for producing light fractions of oil in a process known as "hydro-cracking", refining oil sands bitumen, and removing sulfur from oil. The required hydrogen is produced from natural gas by a chemical process that produces an unwanted byproduct of almost pure carbon dioxide. This is released to the atmosphere. A program is being investigated to capture the carbon dioxide from hydrogen manufacture near Fort McMurray, Canada's main center for oil sands mining and processing, and convey it by pipeline westward, where suitable geologic formations for deep injection are located (Keith, 2002).

Large-scale carbon dioxide sequestration is taking place in the Weyburn oil field in Saskatchewan. Every day about 6.8 million m^3 (240 million cu. ft) of carbon dioxide arrives via pipeline from a coal-to-syngas plant—which heats coal and captures the gases, largely hydrogen, which can be used as a fuel—in Beulah, North Dakota. The gas is pumped a few thousand meters deep where it mixes with oil in the permeable rock formations and helps it to flow toward extraction wells. Since carbon dioxide injection began in 2002, the production of the oil field has increased by 50 percent. When the oil reservoir is drained, all the oil wells will be capped and the carbon dioxide will be sealed in place (Goodell, 2006).

A variation of the sequestration scheme proposes to inject captured carbon dioxide deep into the ocean. One potential site is the Mediterranean Sea near Gibraltar, where water flows out into the deep Atlantic Ocean (Gribbin, 1986). Environmental assessments of this proposal have not been conducted, but given the enormous quantities of carbon dioxide involved there would likely be serious impacts on marine life.

For carbon dioxide capture and sequestration to significantly help reduce global warming, it will need to be practiced at thousands of coal- and oil-fired electricity plants. Many years of work are required before this method can be applied at even a fraction of this scale.

Nuclear Power

Although opposed by many, nuclear power is an energy source with enormous potential, for it produces no greenhouse gases or other atmospheric emissions. In 2006, about 19 percent of the world's electrical capacity was generated by nuclear reactors, with 442 operating nuclear power reactors located in 36 countries. These reactors have a total electrical capacity of 370,700 MW. (Compare this to the effective 24,000 MW electrical capacity provided by wind power in 2007.) Another 28 reactors with 22,500 MW capacity were under construction, with 62 more reactors on order or planned, representing an additional 68,000 MW of capacity. Although there has been no construction of reactors in North America in recent years, nuclear energy is playing a major role in meeting the fast-growing electrical demand in Asian nations including China, South Korea, and Japan.

The latest generation of reactors is vastly safer than earlier models, using new "passive" designs that automatically shut down the reactor in emergency conditions without the need for pumps or human intervention.

Construction time and capital costs have been reduced so the new nuclear plants are cost-competitive with plants fired by coal or natural gas. Other advanced technologies such as breeder reactors can create more fuel than is used, opening the door to a limitless fuel supply (Tammemagi and Jackson, 2009). In addition to producing electricity, nuclear reactors are well suited to supporting energy-intensive industries like oil production from tar sands, desalination of sea water, or making hydrogen for transportation fuel.

Given the daunting environmental challenges posed by global warming and the ever-growing demand for energy, nuclear power has in recent years been receiving renewed attention. In 2006, new reactors were under construction in Finland, Japan, Argentina, China, and India.

In a report issued in 2006, the International Energy Agency, the world's top energy think tank, stated, "The threat to the world's energy security is real and growing". For the first time in its 32-year history the agency recommended the widespread adoption of nuclear power as a clean and secure alternative for electricity generation (Simpson, 2006).

Furthermore, many environmentalists are now supporting nuclear power. For example, James Lovelock (see the text box on page 151) is a strong advocate of nuclear power. He believes it is the only method that can produce clean energy in the large quantities the world needs.

Afforestation

Global warming could be mitigated to some degree by afforestation, that is, the planting of trees, which would absorb carbon dioxide through photosynthesis. The quantities required to make a meaningful impact, however, are enormous. For example, it has been calculated that if an area the size of Ireland (about 100,000 km^2 or 38,600 sq. mi.) were planted with trees each year for the next 40 years, by the time they mature, they would absorb 25–50 billion tonnes (28–55 billion tons) of carbon—the equivalent of 5–10 percent of fossil-fuel emissions. Furthermore, afforestation and other carbon sinks do not absorb any of the pollutants from fossil fuels such as sulfur dioxide, nitrogen oxides (NOx), mercury, and particulate matter.

TREE FACTS

- Twelve trees will over their lives absorb 1.9 tonnes (2.1 tons) of carbon, the amount emitted by an SUV driving 20,000 km (12,400 mi.).
- Trees purify the air by absorbing carbon dioxide, carbon monoxide, ozone, and other pollutants.
- Trees reduce run-off and keep streams clean.

Hydrogen Fuel

Since over 50 percent of oil is used in the transportation sector as gasoline, diesel fuel, and jet fuel, a solution to global warming needs to include the development of portable energy sources that do not emit greenhouse gases. Two avenues are currently being explored: battery-powered vehicles (see Chapter 5) and hydrogen fuel cells. First built in 1830 by Sir William Grove, hydrogen fuel cell technology was later advanced by NASA for use on satellites and space missions. These cells create electricity by joining hydrogen and oxygen together to create water and electricity. There are no other byproducts or emissions. This process is the reverse of electrolysis, where an electric current is used to split water into hydrogen and oxygen. Hydrogen, an easily transportable commodity, can be used as fuel in cars and trucks, and in fact is already being used for that purpose on an experimental basis.

The major drawback is that hydrogen as a fuel needs to be manufactured, and this requires energy. If the hydrogen were produced using coal-fired stations, then the carbon dioxide emissions are merely transferred from vehicle tailpipes to power-station smokestacks. To be effective, hydrogen should be generated by renewable or non-carbon-emitting energy sources, and as we have seen, these are in short supply. The everyday use of hydrogen is still far in the future.

Where Are We Headed?

The United Nations, European Union, and a growing series of studies state that the global temperature increase should be kept below 2 C° (3.6 F°). This is the tipping point that will launch severe climate disruptions that could well become an ecological crisis (Goodell, 2006; Hansen, 2008).

The global temperature has already climbed 0.7 C° (1.3 F°) and another 0.6 C° (1.1 F°) is "committed" or unavoidable given the existing energy infrastructure. In light of the international paralysis in curbing green-house gas emissions, a further temperature increase of 2 C° (3.6 F°) also seems unavoidable. The picture is even bleaker by 2100 when global temperatures will have risen by about 4 C° (7.2 F°). What awaits then?

Perhaps we should heed the words of Stephen Hawking, the renowned cosmologist and mathematician, who says, "Global warming could kill millions. We should have a war on global warming rather than the war on terror" (Earthdive, 2007).

More Signs That the Canary Is Choking

Chased away by the heavy smog and humidity of southern Ontario, my wife and I moved to the west coast of British Columbia in 2005, where we took up a new life deep in an archipelago of dark-forested islands. Perched on a hillside, our house offers a sweeping view; I often sit on a little deck just outside my garret office and watch cloud formations rolling past. Perhaps it's a sign of my age, but I find clouds endlessly fascinating. Frequently, thin layers of cirrus clouds drift past high up, delicate as ferns against the deep blue sky. Occasionally, cumulus clouds thrust upward forming towering, intimidating thunderheads. And at other times as the sun drops low on the horizon, stratus clouds are bathed in the red, orange, and vermilion hues of sunset.

I struggle with the incongruity. This backdrop of ethereal beauty masks the frightening reality that the atmosphere is under an invisible and relentless attack by carbon dioxide, sulfur dioxide, halocarbons, and a variety of other dangerous substances created by the machinery of humans. And the attack has reached epic proportions. Depletion of stratospheric ozone and global warming are unique: they are problems on a global scale, clear signals screaming out that human numbers and activities have reached too high a level.

During most of history the Earth has provided humans with a plentiful supply of food, shelter, and resources. Of course I recognize that periodically over-farming, over-timbering, or other mismanagement have caused localized shortages. But on the large scale, Earth has provided a cornucopia of plenty, allowing humans to prosper and multiply. Now, for the first time, we are bumping against the limits to this bounty. The atmosphere, the weakest, most ephemeral part of the biosphere—the canary in the coal mine of Earth—is gasping.

Global warming, stratospheric ozone depletion, acid rain, and urban smog are at the forefront of the attack on the atmosphere. But there are many other ominous signs that the wastes created by humans have become so massive they are reaching into every corner of the atmosphere. Consider that increasingly there is no place where humans (or other living things) can seek refuge. This chapter looks at some of these lesser known, but far-reaching, problems.

Long-Distance Toxics

In Chapter 3 we saw how Hadley cells and Rossby waves carry heat and air in large global circulation patterns. It is no surprise, then, that toxic substances are carried along with these air currents, traveling long distances to take pollution elsewhere on the planet. Long-range pollution, also known as "transboundary pollution", has been recognized since the 1960s.

The United Nations has had an interest in this issue since 1979 when it established the Convention on Long-Range Transboundary Air Pollution through the UN Economic Commission of Europe (Economic Commission for Europe, 1999). Their observations and model predictions show the potential for the intercontinental transport of ozone and its precursors—fine particles, acidifying substances, mercury, and persistent organic pollutants. Their studies were expanded to include pollutant transport in the northern hemisphere with a view to better understanding air pollution problems in cities as well as their impact on remote areas.

Satellite observations are helping to understand just how far pollutants can travel. The MODIS (Moderate-Resolution Imaging Spectroradiometer) instrument package, for example, is carried aboard the Terra and Aqua satellites (launched by NASA in 1999 and 2002, respectively). This complex of instruments provides information on aerosols and various cloud parameters such as cover, droplet size, and water vapor. MODIS images the entire Earth every one to two days, allowing aerosol trajectories to be plotted and helping to assess and predict air quality (NASA).

In the mid-1990s, scientists at the University of California's Pacific Rim Aerosol Network discovered that pollution crossing the Pacific from Asia was worse than predicted. Every spring, massive dust storms in Asia transport soil eastward to Japan and across the Pacific to the United States. Now researchers have found that sulfate and organic aerosols are also present in the atmosphere (Cahill, 2000). They have discovered that on most spring or summer days, almost one-third of the air pollution high over coastal California cities can be traced directly to Asia. Satellite measurements revealed that high-altitude storm clouds over the northern Pacific have increased up to 50 percent over the last 20 years as the rapid industrialization of China and India has emitted growing amounts of pollutants into the atmosphere.

The researches also concluded that the more energetic Pacific storms could be carrying warmer air and more black soot farther north into the Canadian Arctic where it may accelerate the melting of polar ice packs (Hotz, 2007).

In the mid-1980s, PCBs (polychlorinated biphenyls) and dioxins were discovered in the blood and breast milk of Inuit mothers in the Canadian north—an area that has no factories or other sources of these chemicals. More recently, studies have detected changes in the nervous systems and behavior of Inuit babies that appear to be due to mercury and PCB con-tamination. Surprisingly, in some cases the concentrations of these per-sistent organic pollutants were four to five times the amounts found in women who live in far more industrial southern Canada.

Industrial toxics from hundreds and thousands of kilometers away are being deposited in the Arctic, one of the last wilderness areas of the world. The Inuit are particularly vulnerable because pollutants work up the food chain to accumulate in the fat of fish and game on which much of their traditional diet is based.

Considerable research is directed at understanding the mechanisms caus-ing Arctic contamination. Studies by Julia Lu of Ryerson University (Elliott, 2003), for example, provide some insight into mercury contamination. She found that after the polar sunrise, mercury is converted from a gas into a solid by sun-induced reactions in the frigid polar atmosphere. This aerosol mercury falls to the surface where it accumulates in the ice, snow, and spring meltwater in a form that is easily taken up by the food chain. These mer-cury showers occur each spring just as the ecosystem is awakening and is most vulnerable. Mammals and fish ingest the mercury; it accumulates in animal tissue as it moves up the food chain toward humans.

A study carried out for the North American Commission for Environmental Cooperation modeled dioxin emissions from across the continent and showed that dioxins are transported by prevailing air cur-rents from sources far to the south and deposited in Nunavut, the new Canadian territory in the eastern Arctic (Commoner et al., 2000). Although there are no significant sources of dioxins in Nunavut, or even within 500 km (310 mi.) of its borders, concentrations of dioxins in Inuit mothers' milk is twice that observed in southern Quebec. The study, which included over 44,000 dioxin sources in Canada, the United States, and Mexico, showed the greatest quantity of dioxin in Nunavut (up to 82 percent) comes primarily from the eastern United States. Most of the sources are more than 3,000 km (1,860 mi.) away. Not only is a pristine area of the planet being contaminated, but with the cause lying in a different country, the local government is helpless and unable to pass environmental regulations to protect its own people.

Studies have shed light on how the transport mechanism operates. Persistent organic pollutants—from fossil-fuel power plants, metal-ore smelters, pesticide sprayings, and industry—travel great distances through a series of cycles or hops where they are deposited into water bodies and then re-volatilize back into the air. This is called the "grasshopper effect" or "global distillation" and drives the pollutants to higher latitudes and altitudes where the temperature is colder and the evaporation rates are lower (Max Planck Institut, 2005).

Recognizing that a solution to this problem requires an effort by all of the world's countries, the United Nations organized the Stockholm Convention on Persistent Organic Pollutants, which was signed by 127 countries in 2001. This global agreement will reduce or eliminate 12 of the most hazardous persistent organic pollutants including PCBs, DDT, dioxins, and furans. The signatory nations have developed programs for meeting the objectives and international meetings are held regularly to review progress.

The Great Lakes and surrounding watersheds also receive contaminants from distant places. Scientists noticed that the concentrations of many contaminants were not declining in the way that was expected, given that they are no longer used in the region. This was first noticed when chlorinated pesticides such as toxaphenes, used to spray cotton fields in the southern United States, were discovered in Lake Superior (Brown, 1987). The use of toxaphene was discontinued in North America in the early 1980s, but the compound is still present in the soils of cotton fields in the United States and continues to be transported to the Great Lakes, although in diminishing quantities. Lindane, a pesticide used in Canadian prairie canola fields, is also present; this arrives in larger concentrations in the spring when prairie farmers plough their fields (Walters, 2003).

The Integrated Atmospheric Deposition Network was established in 1990 by the US Environmental Protection Agency and Environment Canada to measure how toxic contaminants are deposited from the air into the Great Lakes. A series of stations at remote locations on each lake gather air and rain samples. They have shown that persistent organic pollutants such as PCBs, PAHs (polyaromatic hydrocarbons), chlorinated pesticides, and trace metals including lead and cadmium are still present in the Great Lakes although they have long been banned in Canada and the United States. Scientists attribute the cause to long-range atmospheric transport from countries where these pollutants are still in use.

A concern is that new chemicals are being detected whose environmental impacts are not yet understood. For example, since the 1990s, chemicals called "polybrominated diphenyl ethers" have been used as a fire retardant on foam padding in car seats and office chairs. They are now detected in the environment in levels of parts per billion, but the concentrations are increasing (Walters, 2003). This is just one of hundreds of new chemicals being brought to market each year without a thorough understanding of how they will interact with the biosphere. Our regulatory agencies are not heeding the hard lessons delivered by CFCs, thalidomide, PCBs, and many other synthetic chemicals.

The studies in the Arctic and the Great Lakes provide a graphic illustration that the human footprint has grown so large that our actions in one part of the globe are having serious repercussions in other parts.

Arctic Haze
In addition to persistent organic pollutants that travel invisibly, the Arctic also has air pollution that is easy to see: a smog-like "Arctic haze". This condition of reduced visibility was first encountered in the 1950s by aircraft flying over Arctic regions. The haze can extend to a height of about 10 km (6 mi.) and has no distinct upper or lower boundaries. It appears blue-gray when viewed away from the sun, and reddish-brown when looking toward it. The haze is a modern phenomenon; we know that in the 1800s Arctic visibility was clear and sparkling with mountains 200 km (120 mi.) away clearly visible. Today when the haze has settled in, visibility is often less than 35 km (22 mi.) (Canadian Encyclopedia, 2008).

The haze is seasonal, lasting from approximately February to May and peaking in the spring. It is most intense and persistent when stable, high-pressure systems produce clear, calm weather. The haze forms patches 800–1,300 km^2 (300–500 sq. mi.) large, and often appears in distinct bands at different heights, since the warmer "dirty" air is forced upward over the dome of cold Arctic air.

With no industrial sources of pollution in the Arctic, many natural

. causes were proposed for the haze. It was not until the 1970s that the haze was found to result from human activity. Using aircraft equipped with lasers and high-volume vacuum air samplers, scientists found that the haze is Arctic-wide, covering almost all of the area north of 60° latitude. The haze is composed of a range of compounds including sulfur dioxide, soot, and hydrocarbons, as well as natural oceanic and soil materials. In the spring the Arctic tilts toward the sun, and sunlight triggers chemical reactions that change sulfur compounds and other pollutants from gases to microscopic solids and liquids.

Tracing the origin of the pollutants was difficult, given that in the 1970s there was little meteorological data available for the Arctic. In 1980, scientists developed a method of measuring the isotopes of air pollutants that allowed them to trace their source back to the region—even to the very factory—that created the pollution. They discovered that about 66 percent of the haze comes from industrial regions in eastern Europe and western Russia with most of the remainder coming from western Europe. Surprisingly, only about 4 percent originates in North America, because prevailing wind patterns carry North American air pollution eastward over the Atlantic where it is dispersed by storms.

The other mystery surrounding Arctic haze was its occurrence in the winter and spring. Studies showed that the Arctic atmosphere absorbs large loads of pollutants in the fall and winter because a stable temperature inversion prevents vertical mixing. Prevailing winds generally flow northward during this period (in the summer, they flow southward), and particles stay in the air longer because there is no rain and little snow to wash them out.

Since the haze traps radiation and soot makes snow darker, the heat absorbed by the Arctic may be increasing. This elevation of temperature, of course, affects the regional climate. Sulfur dioxide also makes the ice and snow more acidic. The acidity can have adverse environmental effects, especially during the spring melt (Pollution Probe, 1991). Much research remains to be done to determine the health effects of Arctic haze on the ecosystem and human well-being.

The news is not all bad. Scientists have determined that the haze has stopped growing. Furthermore, although the level of sulfate—the main contaminant—has not changed, the concentrations of other pollutants in the Arctic have decreased in recent years. Lead, for example, has declined by 5 percent since 1980, mirroring the phase-out of lead in gasoline. Levels of the insecticide lindane have also dropped significantly.

These decreases indicate that international efforts to phase out and ban toxic substances are working. Considerable progress is still required, however, before distant mountain peaks will once again sparkle on the Arctic horizon.

Asian Brown Cloud

Although smog is usually associated with big cities, a large brownish pollution haze often covers the whole northern Indian Ocean and much of India, Pakistan, Southeast Asia, and China. The haze layer lasts for three or four months each year and is facilitated by an extended dry season that prevents the removal of pollution by rainfall. The cloud peaks during the winter, when the temperature of the air tends to be hotter than that of water and ground, creating an inversion that traps the pollutants. Researchers found the haze was made of soot, ash, dust, and airborne chemicals—all products of man-made pollution.

Scientists warned that the cloud, estimated to be 3 km (1.2 mi.) thick, is responsible for hundreds of thousands of deaths a year from respiratory disease. By slashing the sunlight that reaches the ground by 10–15 percent, the choking smog has also altered the region's climate by cooling the ground while heating the atmosphere (Ramanathan et al., 2007). This "regional cooling" may slightly reduce the impact of global warming. Also, when the amount of sunlight is reduced, the evaporation of ocean water is reduced. This may have a significant impact on diminishing water supplies in the region.

Since this discovery, the impacts of aerosol pollution around the world have received considerable attention. Exactly how much aerosols might counteract global warming is still under investigation.

Hazy Visibility

Not long ago the health impacts of polluted air were often overlooked. It was felt that such pollution was unavoidable and it was invisible—out of sight and out of mind. Urban smog changed that mindset and now it is recognized that the brown haze hanging over cities has serious health implications. In recent years smog has reached much further afield, becoming a serious problem even in areas where no pollution is made. Arctic haze and the Asian brown cloud are examples of this phenomenon taking place with relatively well-defined boundaries. But regional haze is also becoming more prevalent as a general problem without distinct boundaries.

If you think you can escape the city to enjoy clean air in the wilderness, think again. Over the last ten years, the average ozone levels in 29 US national parks rose by over 4 percent. Regional haze is reducing our enjoyment of what is considered a personal freedom: the ability to enjoy nature and views of mountains against a cloudless deep blue sky. In Virginia, Shenandoah National Park's annual average visibility has decreased from approximately 145 km (90 mi.) to less than 30 km (19 mi.), and is often much lower in the summer. Pollution is frequently so bad that it obscures the majestic views that gave Blue Ridge its name. Mammoth Cave National Park in Kentucky has the poorest visibility of any national park in the United States due to its close proximity to some of the nation's most polluting coal-fired plants in the nearby Ohio Valley (National Park Services, 2002). And distressingly, on some of the more severely polluted days in the Grand Canyon, the North Rim is hardly visible from the South Rim— only 15 km (9 mi.) away (National Park Services, 2002).

Some parks in the United States, including the Great Smoky Mountains National Park in Tennessee, Acadia National Park in Maine, and Shenandoah in Virginia, have had to issue health warnings because of smog. In Acadia some smog episodes have been worse than the pollution in the cities of Boston or Philadelphia. The Great Smokies, the most visited national park, issued more than 100 unhealthy-air alerts between 1999 and 2001 (Seelye, 2001).

Haze obscures the color, clarity, texture, and form of what is seen by the human eye. Some haze-causing pollutants, primarily fine particles, are emitted into the atmosphere by electric power generation, truck and auto emissions, forest fires, and construction. Others are formed when gases emitted into the air form particles. Examples include sulfates formed from sulfur dioxide and nitrates from nitrogen oxides (NOx). Because winds can carry these emissions right across an entire continent, haze can occur anywhere in North America.

By the mid-1990s regional haze in the United States had become such a nuisance that action was required. An amendment to the US Clean Air Act in 1997 called for rules to improve visibility in parks by 15 percent per decade with the goal of achieving pristine air quality by 2064. The rules apply to power plants and call for state and federal agencies to work together to improve visibility in 156 national parks and wilderness areas such as the Grand Canyon, Yosemite, the Great Smokies, and Shenandoah.

Radioactivity in the Atmosphere

When ranking the most stupid things humans have ever done, the testing of nuclear weapons in the atmosphere places very close to the top.

From 1945 to 1963, the Soviet Union, the United States, Great Britain, and France developed nuclear weapons and tested them with above-ground detonations. During the Cold War bomb race, the United States conducted 216 nuclear tests and the Soviet Union, France, and Britain detonated an additional 150 (Sherman, 2004). The heat of these explosions was so great that the blast products were carried high into the troposphere and even penetrated into the stratosphere. The radioactive materials were then dispersed by winds and carried around the globe. A more effective way of polluting the atmosphere could not have been devised.

The radioactive elements of principal biological concern are shown in the following table.

Element	Half-Life
tritium	12 years
strontium-90	28 years
cesium-137	30 years
iodine-131	8 days
manganese-54	314 days
iron-55	2.7 years
cobalt-60	5.3 years

The term "half-life" refers to the amount of time needed for a quantity of a radioactive element to decay to half that amount. It can be used to measure how long the radioactivity will remain in the atmosphere. By the late 1950s, the increase in atmospheric radiation was measurable around the world and had become a serious health concern.

On August 5, 1963, the United States, the Soviet Union, and Great Britain signed the Treaty Banning Nuclear Weapon Tests in the Atmosphere, in

Outer Space, and Under Water, known more familiarly as the Limited Nuclear Test Ban Treaty. Atmospheric bomb testing was banned and subsequent tests were conducted underground.

In 1989 the collapse of the Soviet Union and the end of the Cold War signaled another significant and very welcome movement away from nuclear weapons. This included a decrease in the testing of nuclear weapons and also in the size of nuclear arsenals. The Non-Proliferation Treaty was extended for an indefinite period and a Comprehensive Test Ban Treaty was recently adopted by the United Nations. Today, even underground blasts are banned by international treaty and nuclear bomb testing is done by computer simulation. Tests in the Pacific and elsewhere have ceased. South America and Africa are continents free of nuclear weapons, and the United States and the Russian Federation are downsizing their nuclear arsenals. Running counter to this trend, unfortunately, are the underground bomb tests conducted by India and Pakistan in 1998. The nuclear programs of Iran and North Korea are also of grave concern. Fortunately, radioactivity decays over time so the atmosphere has made an almost complete recovery from the earlier tests.

When someone is exposed to radiation, the dose is measured in units called microSieverts. The dose received by a typical person from atmospheric weapons testing has decreased dramatically from a maximum of about 150 microSieverts per year in 1963 (when atmospheric testing ceased) to less than 5 microSieverts per year in 2002. The latter represents a trivial contribution (less than one-quarter of 1 percent) to natural radiation from the surrounding environment (Tammemagi and Jackson, 2009).

Chernobyl

The notorious explosion of the Chernobyl nuclear reactor in Ukraine on April 26, 1986, released large amounts of radioactivity into the atmosphere. Air currents carried the cloud in two directions, reaching North America's east coast on May 6 and its west coast on May 7. The highest air concentrations in Canada occurred in May and by the end of June the levels had returned to normal. Health Canada estimated that the Chernobyl accident led to the typical Canadian receiving a radiation dose of 0.28 microSieverts, a negligible amount compared to typical variations in natural background or to dental or chest X-rays.

The Oceans

Airborne pollution not only despoils the atmosphere and the air we breathe, but also affects another vital global system, the oceans. In recent decades scientists have discovered a number of large "dead zones" in oceans, which shrink and swell, pulsating with the rhythm of the seasons. Their causes are varied: massive runoff of fertilizer from farm fields; extreme rainfall that flushes more pollutants into water bodies than normal; extreme drought that reduces the inflow of water, concentrating harmful substances in water bodies; and ocean current shifts suspected to be linked to global climate change. These dead zones are sharply depleted in oxygen, removing their ability to support life. These areas can persist for weeks or months. The number of dead zones around the world is increasing rapidly, rising from about 75 sites in 1990 to about 150 in 2004; and, alarmingly, they continue to grow. Often occurring near the deltas of major rivers, the dead zones are said by the United Nations to pose as big a threat to fish stocks as overfishing (*Sidney Morning Herald*, 2004). One of the world's largest dead zones occurs each year in the Gulf of Mexico off the Texas and Louisiana coasts. In 2007, it stretched to more than 22,000 km^2 (8,500 sq. mi.). At about the size of New Jersey, it was one of the largest encountered since measurements began in 1985. Furthermore, it has been convincingly linked to massive deaths of fish and shellfish (NOAA, 2007).

In the 1980s, investigations of drill cores representing over a century of ice accumulation showed that atmospheric acidity levels were relatively constant in the early 1900s. This situation changed significantly after 1956, when acidity increased by 75 percent over the next 25 years. This increase correlates with the doubling of industrial sulfur dioxide emissions worldwide. Since the mid-1950s the burning of coal and the other fossil fuels has accelerated. As the concentrations of carbon dioxide and sulfur dioxide increase in the atmosphere, the oceans—covering 70 percent of the planet's surface—absorb a fraction of these compounds. Once in the water the gases are transformed into carbonic acid and sulfuric acid. This, of course, raises the acidity of the oceans. In turn, this increased acidity, combined with rising sea temperatures, contributes to the rapid decline of the world's coral reefs. It is also damaging fish and other marine life.

The two lungs of the world, the atmosphere and the oceans, are closely linked. When immense and vital worldwide systems such as these are suffering from human activities, it is urgent that we take action. But how can we proceed against such an immense problem? What can we do? The next chapter discusses some potential ways to move forward.

The Path Forward

To step back and look at the planet from space, it changes your values,
your perspective on things ... Most astronauts come home from a mission
with a deeper understanding, deeper commitment to our planet Earth.
—Roberta Bondar, the first Canadian woman in space

Nothing shouts the message that the natural systems of our planet are being overwhelmed more loudly and clearly than the degradation of our atmosphere. At first, air pollution problems were localized, adjacent to areas of industrial activity; by and large, the atmosphere, with its self-cleansing mechanisms and immense capacity to dilute, kept pace. But now, for the first time in history, the human pollution has become so vast it is overwhelming the defense systems of the entire globe. Ozone depletion and global warming are unassailable signs that we are pushing against the limits of our beautiful planet. In this chapter we step back and, like the astronauts in space, take a big-picture view of the environment and the state of the world.

A Society Heading for Collapse?

A question has long worried me: what lies beyond global warming? The human population, along with our livestock and fish farms, is increasingly crowding the planet. Resources are running out, pollution is mounting, and new viruses and bacteria proliferate. At the same time, scientists and engineers continue to develop powerful new technologies and a bewildering array of synthetic chemicals. It is inevitable that more complex problems will surface to challenge our ingenuity. Global warming presents an almost unsolvable challenge. How will we ever cope with what lies beyond?

Perhaps I'm being overly pessimistic, or perhaps my concerns are prescient and valid. Whatever the case, whenever my mind turns to global warming—which it does all too often—I have a foreboding that the Earth's rising temperature marks the beginning of the end of modern civilization.

The long road of history is littered with the carcasses of civilizations that have soared to great heights, and then suffered rapid and severe crashes. The Sumerian, Egyptian, Roman, Greek, Mayan, and Aztec empires come to mind. The decline of a smaller civilization, Easter Island several centuries ago, resonates with our own complex situation today (Diamond, 2005).

Here was a delightful south Pacific Island, with palm trees that swayed in the breeze, provided shade from the heat of the sun, and were a reliable and renewable source of wood and leaves for construction of homes and an array of items useful for everyday life. In particular, the palm trees provided the raw materials for boats, which allowed the people to gather a rich harvest of seafood provided by the ocean. It is little wonder that once this bountiful little island was inhabited by a handful of Polynesians, who arrived in about 500 CE, and society thrived.

But something went wrong, horribly wrong. The population grew and grew, reaching about 10,000 by about the year 1100, placing an enormous stress on the small landmass of 166 km^2 (64 sq. mi.). Its society, which was organized into clans, competed in raising ever-larger stone statues, called *moai*, to honor their ancestors. Cutting, moving, and erecting the stone gods required rope, strong wooden beams, and rollers. This required cutting down palm trees. As competition increased, more and more palms were toppled to satisfy the ever larger and more extravagant statues. Instead of protecting and nourishing the thing that was most essential to their existence, the people consumed it in their self-destructive obsession with stone monuments. Instead of preaching restraint, the nobles and priests urged them on. Trees were cut faster than they could grow until, finally, in about 1400 CE the last tree fell.

For another generation there was enough old wood on the island to build boats and haul stones for more statues. By this time about 800–1000 moai had been erected. But the time came when even the old wood was gone, and then a bad situation turned worse. Without trees, rain washed away the rich volcanic soil; and of course boats could no longer be built for fishing. The warrior class seized power and about 90 percent of the population perished in an orgy of violence and cannibalism. In a final convulsive fit, violence erupted again in the mid-1700s and, in addition to killings and atrocities, all of the statues were toppled.

When I first heard this story I was appalled at the blind stupidity of the islanders. What were they possibly thinking as they cut down that last tree? What kind of pathetic leadership did they have? What happened to accountability, prudence, and planning? Surely, in today's technologically advanced age, where education levels are high, democratic governments flourish, and news and information are readily available, this kind of tragic outcome could never happen. Surely we have learned from the past. Haven't we?

But uncertainty keeps nibbling at the edge of my mind. There are too many parallels with Easter Island to allow me to sleep well at night or to hug my granddaughter with any sense of optimism. Now the globe has become a small place, and just like on Easter Island, the population has increased beyond what can be sustainably supported. The United Nations, scientists, and other respected organizations tell us we are consuming resources—fossil fuels, fisheries, water—faster than they can be replaced. Nevertheless, we continue to look for "economic growth". Our leaders do not preach restraint nor do they plan for the future; instead they look for increasing economic productivity and consumption of material goods. Technology, they claim, will find solutions for the future.

I was reminded of Easter Island on a recent holiday to Phoenix in southern Arizona. I would have thought this starkly beautiful yet inhospitable desert landscape would inhibit settlement and encourage a Spartan lifestyle. Instead, the opposite is true. Phoenix is the fastest growing metropolitan area in North America. Its population more than doubled from 1.5 million in 1980 to 3.7 million in 2005, and the growth continues relentlessly with a projected population of over 15 million by 2050.

The lifestyle in greater Phoenix does not match its austere surroundings: rather it is obscenely lavish and revolves around materialism. The automobile rules; broad, multilane roadways run everywhere. There is no meaningful public transport and the absurdly oversized Hummer is a common sight. Smog over the city is common. Rather than the saguaro cactus, a more appropriate icon for this region would be a gas-guzzling SUV or pickup truck.

Water, of course, is scarce, and is an absolute necessity for life in the desert. Yet it is consumed as if there is no tomorrow. Water pumped from aquifers has caused the water table in some areas to drop by 60 m (196.9 ft); water piped from the Colorado River has helped its once-mighty delta turn to a trickle in summer.

In the searing desert temperatures, every home and every car is equipped with air conditioning; 68 percent of this energy-intensive equipment is powered by fossil fuels. As the population soars, the additional electricity demand will be met almost exclusively by fossil fuels, particularly by coal. As on Easter Island, Phoenix's leaders do not preach prudence or conservation. There was, for example, no recycling at our resort; nor was there any location to which we could deliver our recyclables. Local politicians vociferously promote further growth. Meanwhile, the veil of smog over Phoenix thickens and the temperature rises.

The Enemy Lies Within

Although I despair, I also seek hope. Surely there is still time and resolve to rescue the future. How do we proceed?

The first step is to recognize that relying solely on future scientific and engineering breakthroughs will not suffice. Science has brought wonderful advances that have enriched our lives. But our clumsy, almost childlike use of powerful technologies has also brought enormous problems like the ozone hole, nuclear wastes, thalidomide, acid rain, childhood cancers, and more.

We must use the solutions technology has provided us. Yet hybrid cars, smokestack scrubbers, carbon sequestration, and solar and wind power simply are not enough. As for the automobile, as we saw in Chapter 5, these green technologies will not be able to keep up with the population-economy treadmill.

To make progress in reducing global warming we must slow this treadmill. In other words, we need to slow down economic and population growth.

It will not be easy. There are strong forces at play, and societies have enormous momentum; like an ocean tanker, social directions cannot be changed quickly. Powerful corporations and organizations—think here of the priesthood of Easter Island—have a vested interest in maintaining the status quo. Furthermore, many people are frightened of change and the uncertainty that comes with it. We have already seen the enormous opposition to the Kyoto Protocol. And, of course, very few people will happily reduce their standard of living. Furthermore, as evidenced by declining fish stocks in international waters, humans do not always act well collectively (that is, at the international level). Instead, we seem driven by tribalism and self-interest.

So how do we move forward? Let us look at population first.

A Very Crowded Planet

Starting in about 1800 human numbers suddenly soared, accelerating from about one billion to the current 6.5 billion in just over two centuries. Today, the population continues to increase robustly, although it is projected to slow and stabilize at between nine and ten billion in the second half of the twenty-first century. It seems clear that the numbers present today—without further growth thrown into the equation—exceed what the planet can support in the long term.

Very little is being done to address human population. It is a topic that for many is enshrouded with religious, ethical, and even fanatical taboos. First and foremost, the population issue must be brought out of the shadows and discussed openly, frankly, and constructively. Issues such as family planning, family allowance, birth control, immigration, and other population-related policies must be debated and, furthermore, addressed in an integrated manner together with environmental and economic strategy.

Slowing and halting population growth should become an important goal. Every nation should have a population policy, which is overseen at the cabinet level, and coordinated internationally through the United Nations (or a similar organization).

FIGURE 11-1

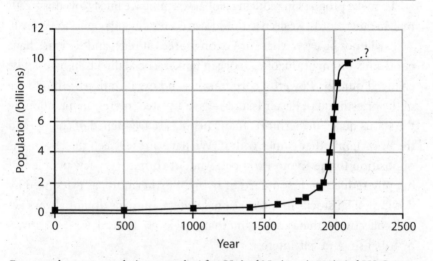

Runaway human population growth. After United Nations (1999) and US Census Bureau (2004)

The issue, of course, is most urgent in highly populated, developing countries such as India, Pakistan, China, and Indonesia, where 95 percent of future growth will take place. Nevertheless, population must also be addressed in North America where it continues to increase at a rate of about 1.1 percent per year, about two-thirds of which comes from immigration. The United States, in particular, as the third-most populated country in the world and with the greatest per-capita consumption, must rein in its population growth.

Rich nations face a difficult challenge for they must control population growth, reduce their levels of consumption, and at the same time, help the poorer, developing nations. Many people, including former US Vice President Al Gore, believe that addressing the huge disparity between poor and rich nations is the key to creating a globally sustainable environment. Studies have shown that only after sufficient wealth has been accumulated can nations afford the cost of environmental protection. Before that, the simple problems of ensuring food, cleanliness, and security come at the top of the agenda.

With an increased standard of living comes improved education. Studies show that education and high literacy rates, especially for women, are key factors in curbing population growth. Education also leads to a more satisfying quality of life, including a knowledge and appreciation of our environment.

Managing Consumption

The total demand on the Earth's resources (sometimes called "the human footprint") is directly proportional to how many of us there are and how much each of us consumes. One cannot be considered without the other. As the standard of living, driven by access to cheap energy, has increased so has our desire for consumer goods. The average US family owns 2.2 cars (Mintel, 2005). Most couples aspire to a sprawling home and, if possible, a vacation home. And, of course, everyone has to have a cell phone, digital camera, DVD player, and the latest desktop and portable computers accompanied by flat-screen monitors, color printers, and Dolby speakers. And the economy keeps on growing and churning out these objects.

Many thinkers feel that technology is at odds with environmental protection. It is obvious that our increasingly technological (and urbanized) world leads to a separation from the environment and a dwindling recognition of our place within a complex natural web. But technology is not to blame. It is our materialism that results in millions and millions of disposable products whose manufacture and disposal saps the world's resources.

Nowhere has consumption reached a higher fervor than in the United States. As the world leader in scientific and technological innovation, the United States has amassed enormous wealth which is shared widely by its citizens. Fuelled by economic prosperity and low energy prices, consumption has increased steadily over the past 50 years. The United States and Canada have the highest per capita consumption in the world as measured by the gross national product (GNP; an indicator of the size of the economy). Yet much of this profligate materialism could easily be reduced without sacrificing the quality of life.

North Americans are creating a major share of the world's pollution. With only five percent of the world's population, North America (primarily the United States) generates some 21 percent of the world's carbon dioxide emissions (EIA, 2005). It is similar for nitrogen oxides (NOx), sulfur dioxide, and other emissions. And these totals do not even reflect the complete picture: not included here are the emissions from overseas factories that make goods for North American consumption.

The United Nations Environmental Program calculates that the typical North American has an environmental footprint four times bigger than the world average (United Nations, 2002), and 30–50 times greater than that of the average person living in India (United Nations, 1998). Significantly, the UN concluded that such a rate of consumption is not environmentally sustainable. Yet as a society, we don't seem to care.

Although population has grown rapidly in the past century, the economy has expanded even faster. Since the early 1900s the world's population increased four-fold, but the economy positively rocketed, increasing by forty-fold (Wright, 2004). The economy, as measured by the gross national product, is an excellent indicator of the stress we are placing on the planet. The GNP, rather perversely, increases when the environment is damaged. For example since the GNP includes all monetary transactions, the cost of cleaning up a major oil spill increases the GNP.

The economy is one of the main forces preventing us from taking corrective action against environmental problems, for many economists preach there are neither limitations nor constraints to ever-continuing growth.

THE PATH FORWARD / 225

Very few politicians or other leaders want to slow the economy. They believe this will bring job losses, reduced corporate profits, a diminished tax base, and decreasing numbers of construction and other projects.

That economic growth should have priority over the environment is neither ethical nor logical, yet it dominates political thinking. John Howard, former prime minister of Australia, summed it up nicely: "What we have to do is to find ways of reducing greenhouse gas emissions that do not cost us an economic price" (Woods, 2006). Canada's prime minister in 2007, Stephen Harper, mouthed virtually the same words as he pledged to withdraw Canada from Kyoto. And of course, George W. Bush, America's president from 2000 to 2008, refused to sign the Kyoto Protocol for exactly this reason.

All economic (and political) thinking revolves around a growing GNP, and if the GNP growth falls below even two percent annually, it is considered "stagnant".

Economic growth, however, requires a growing population to supply more workers and more consumers. As long as human numbers are increasing, there will be a need for more roads, buildings, and factories. Growth needs more people, and people enjoy the prosperity brought by growth. The economy and population are like two oxen yoked together, always pulling forward.

There are many resemblances of the mantra of "endless growth" to so-called pyramid schemes, which work on a similar principle. This kind of scheme requires an expanding number of people to feed those already in the game. We all know how a pyramid scheme will collapse at a given point. The gullible may believe that everyone can win; but the reality is, someone will have to face big losses.

There are numerous signs that we have surpassed the carrying capacity of mother-ship Earth and, just like fruit flies propagating rapidly inside a sealed bottle, we are approaching the endgame. Our refusal to face the problem is caused largely by our affluence. We fear to lose it. But we will lose much more if we don't take action.

Effective leadership is needed, especially here in North America. It is not morally right for North Americans to be consuming the world's resources and degrading the environment disproportionately. We must accept the responsibility for rectifying the situation. We need to reexamine the way we live, our values, and the mad pursuit of materialism. In the past, philosophers and revolutionaries have led us to rethink the basic premises that underpin our society; now it is nature itself calling to us.

A Vision: Equilibrium

The challenge we face is to curtail both the burgeoning population and the soaring economy. How do we achieve this? Where do we start? As with any project, small or large, the first step is to establish a vision, a goal toward which we aspire.

> **IROQUOIS VISION**
>
> We could learn from the Iroquois nation, which requires its tribal councils to formally consider the impact of their decisions on the seventh generation into the future, a period of approximately 150 years (Gore, 1992).

An important initial step has been taken in the right direction. It is a philosophy called sustainable development, introduced by the UN's Commission on Environment and Development in 1987, to guide nations in the conduct of their commercial and industrial activities. The definition is: "Sustainable development is development that meets the needs of the present without compromising the ability of future generations to meet their own needs." It can be viewed as the integration of economic, social, and environmental systems. This is a major change. Usually the environment is either excluded in political and social debate or given much lower priority.

Sustainable development is fine in theory; but does it go far enough, and how can it be put into practice?

A few years ago while hiking along the Niagara Escarpment of southern Ontario I glimpsed the true meaning of sustainable development. I walked in the shade of stately sugar maples, oaks, walnuts, sycamores, and other trees of the Carolinian forest. Their long arms reached toward the sky protecting an understory of dogwoods and staghorn sumacs. Small saplings grew in the sun-dappled shade. Fallen logs lay here and there, covered in green moss, slowly decomposing and forming a habitat for insects, small animals, and birds. Countless generations of trees had fallen and then risen again in an endless cycle on this delightful slope.

A stream bubbled through a ravine lined with cobbles on a meandering journey to Lake Ontario. Puffy cumulus clouds floated in the blue sky, evidence that water was evaporating and rising from the lake into the air. Later in the week the white clouds would turn a sodden gray and rain would descend, soaking the trees and ground where I now stood. The rainwater would tumble down the stream, enter the lake, and again rise to form clouds. I sat on a log and thought about the cycles of water and life that have been repeated here for millions of years.

Nowhere was any part of the hillside in rampant, runaway expansion. No tree, shrub, or animal was expanding voraciously, driving out, and decimating all other species. Instead, everything was in a gentle equilibrium. Every piece had a niche in which it performed a vital role in supporting the system around it. Sometimes a tree topples or a shrub dies; it is replaced by new growth. Only the amounts that are lost are replaced. I saw that nature has designed a system that has worked effectively for hundreds of millions of years. It is equilibrium.

That is the goal we must pursue. We humans need to step away from constant growth and seek equilibrium in both the economy and our population.

Achieving the Goal

Defining the goal is relatively simple. Taking steps to achieve it is far more of a challenge, probably the greatest ever faced by humans. I do not claim to have the answers, for there is no simple panacea. I will, however, make a number of suggestions.

An Environmental Ethic

What is needed most of all is a change in our mindset. We need an ethic where we value the environment, recognize that it is being steadily degraded, and choose to take actions that remedy the situation. We must recognize that humans are an integral part of nature—as it goes, so go we—and we must emulate its grand scheme of equilibrium.

And, yes, sacrifice will be required. In North America, a major step toward containing consumption would be to decrease our per-capita use of energy. This may not be that difficult: other nations such as Switzerland, Germany, Sweden, and Japan use about half the energy per person as North Americans do. Yet they have rich and satisfying lives and do not find it burdensome to live in higher-density housing, drive smaller cars, use mass transit, and provide their own bags when shopping.

There is a significant difference between "standard of living" and "quality of life". The former represents the level of consumer goods we own. The latter includes the quality of air we breathe, the richness of friends and community, and the peace of mind knowing that our grandchildren will enjoy the simple pleasures of wildflowers in a meadow or the graceful leap of a deer.

An environmental ethic must also be developed at the corporate level. A new approach using "environmental management systems" (EMS) is receiving growing attention. An EMS is voluntary, generally goes beyond just meeting regulatory compliance, is integrated with the overall management of the company, and requires the creation of environmental awareness in all employees.

An international standard, ISO 14001, similar to the quality assurance standard ISO 9000, provides a standardized framework for EMSs. Many firms are formally registering their EMSs as ISO 14001-compliant because their customers are demanding environmental responsibility. As ISO 14001 gains popularity around the world, it will encourage countries to harmonize their environmental laws, thereby raising environmental protection

LEADER OF THE CORPORATE GREEN PACK

Robert Schad, CEO of Husky Injection Molding Systems, named "Canada's Most Environmentally Responsible Company" in 2002, explains, "We believe it makes good business sense to integrate sound environmental, social, and economic considerations into all our decisions and practices." In 2001, Husky's net carbon dioxide emissions were 15 percent less than in 1990, far exceeding Kyoto targets. This is remarkable because Husky's operations more than tripled in size over the past 12 years. Husky does not plan to rest on its laurels; it aims to completely eliminate net carbon dioxide emissions by 2010.

Another program to reduce and eliminate waste earned Can$1.2 million over two years by selling excess metals, plastics, and cardboard—previously, the company had paid money to have this waste removed. An energy retrofit program saved more than Can$500,000 in electricity costs at its main facility alone.

Husky encourages its employees to "act green" by offering shares to staff who recycle, plant trees, carpool, buy hybrid vehicles, and volunteer their time to environmental causes.

in countries where it currently lags. EMSs should become a requirement not only for companies, but also for governments.

We can hope that such a groundswell of support for the environment will develop that politicians will have to listen.

Pricing the Environment

Society is driven by economics, and the environment is treated as a minor player. The two need to be brought into harmony.

The cost of environmental damage is treated as largely external to the economy and, therefore, is not reflected in the market price of the various goods and services that cause pollution. We need to recognize that the Earth's natural ecosystems and the "services" they provide have an enormous economic value. Costanza et al. (1997) estimated the 1997 value of the biosphere to be approximately $33 trillion per year, which is almost double the global GNP. Because this value is not included in economic analyses, environmental considerations are not adequately represented in policy decisions.

In fact, it gets worse. Governments often subsidize fossil-fuel-driven projects such as the development of new oil and gas fields, the development of tar sands projects, and the construction of pipelines. In transportation, governments encourage the use of SUVs by allowing them lower gasoline-efficiency requirements than regular cars. The list goes on.

A comprehensive change is needed that properly integrates the price of the environment into all economic transactions. For example, a Canadian government task force on acid rain calculated the health and associated benefits of a 25 percent reduction in sulfur dioxide emissions and determined that savings in avoided health-care costs were about Can$210 million in Canada alone (Tollefson, 2000). Clearly, sulfur dioxide emissions exact a large burden on the health-care system, a cost that polluting industries are not paying. Furthermore, these emissions bear additional costs associated with the acidification of lakes and damage to forests.

One way of factoring environmental costs into the economy is through the credit trading systems discussed in earlier chapters. Some economists believe emission-reduction credits will become the currency of the twenty-first century and companies will allocate them with the same care as they allocate capital funds today.

Harvard professor Michael Porter argues that society should consider pollution as a form of economic waste, a clear sign that resources and energy are being used inefficiently (Tollefson, 2000). By putting a price on the environment we will recognize that we need to preserve its capital and learn to live on the interest alone.

Reform Political Systems

Why do we behave like the Easter Islanders and take no action as resources dwindle, people in sub-Saharan Africa starve, and polar ice cover disappears? Al Gore points out that "… the political system is simply not working", adding, "one of the most deadly threats to the stewardship of democracy is a lack of leadership" (1992).

It is not the role of this book to outline specific reforms to political systems, but I do feel strongly that change is needed. Many North American politicians' primary motivation is to maintain a positive image in the media so they can be reelected. Their actions are driven by personal short-term goals rather than what is good for society in the long term. For them, it is difficult to pass tough laws no matter how desperately needed they might be. And, of course, their situation is not made easier by the media who also have short-term interests and a tendency towards sensationalism rather than objective analysis.

The current democratic government system is vulnerable to the proliferation of organized lobby groups. One of their main tactics is to argue that nothing should be done until the science is better understood. This is in spite of the adoption of the "precautionary principle" at the Earth Summit in Rio in 1992, which states that where there are threats of serious or irreversible damage, a lack of scientific certainty is not a reason for governments to delay taking cost-effective counter measures. Our politicians listen to lobby groups, but they ignore the UN Earth Summit. Why?

The North American political structure needs to be amended so our leaders focus less on media posturing and more on leadership. They need the tools and support for taking tough stances when necessary. There is also a desperate need for personal accountability in the system. How do we achieve this?

Because air pollution drifts across national boundaries, an international body with teeth is needed to enforce environmental standards on a global scale. Serious consideration should be given to Al Gore's recommendation that the United Nations establish a Stewardship Council similar to the Security Council, which would be responsible for Earth's ecosystems (1992).

A serious impasse has developed on the international stage. Poorer countries are moving towards industrialization, and richer countries are unable to rein in their consumption. Meanwhile, little is achieved. The Stewardship Council could play a major role in resolving this impasse.

A CITY FOR THE FUTURE

Sweden is demonstrating that green cities are possible. A former industrial harborfront in Malmö has been transformed into one of the world's most sustainable communities (EAUE, 2004). In the first phase about 1,000 housing units were constructed on a 25 hectare (62 acre) site that includes businesses, schools, restaurants, and services. A nearby medieval town provided inspiration for the design of the attractive futuristic buildings. Located near the center of the city (not in a distant suburb), the development has won awards and has many unique features.

- Priority is given to pedestrians and cyclists with a large network of paths, and cars are limited to parking lots on the outskirts. Public transportation is included as well as a car-pool fleet using electric vehicles and alternative fuels.
- Green spaces are abundant, including a communal park and rooftop gardens. The developer is responsible for planting large trees and their long-term maintenance.
- The development consumes almost 100 percent renewable energy obtained from solar photovoltaic panels, wind turbines, and a heat pump connected to an underlying aquifer.
- A comprehensive recycling system includes gathering organic wastes via a piping system, treating them in an underground tank, and applying a new technology to remove phosphorous.

The Future?

We humans are poisoning the atmosphere, and in the process we are quietly choking both nature and ourselves. We must learn to recognize, on a societal scale, all of the warning signs; and we must realize that we are running out of time.

I don't wish to be overly pessimistic. There are pockets of hope that glimmer like stars against a vast black sky. Progress is being made in areas such as hybrid cars, smart growth, electricity from windmills and photo-voltaic cells, and EMSs. To date, unfortunately, these positive steps are not keeping up with the population-economy treadmill. We must run faster; we must do much more.

It seems incomprehensible that humans have made such enormous strides in science and technology, yet achieved so little progress in address-ing issues of pollution. Universities are full of brilliant people improving the speed of computers, making composite materials stronger, finding ways to transmit messages faster via fiber optics—and fashioning all these advances into an ever growing cornucopia of consumer goods.

We continue to be faced with a series of complicated questions. How can humans live in harmony with nature and the environment around us? How do we modify political systems to promote responsible leader-ship? How can we reduce conflict and increase cooperation between nations? How do we foster individual lifestyles that are environmentally sound and not based on materialism? How do we equitably share wealth around the globe? How do we bring human population to a sustainable level? Society should value and encourage such social developments just as much as it values technological progress. Surely preserving the envi-ronment and our society is worth just as much as building a missile defense system or placing a human being on the moon.

We need to look inside ourselves, for the solution to this dilemma lies in the values we espouse. These values must recognize the critical impor-tance of the environment, and that we humans are not masters of nature, but an integral, functioning part of that intricate scheme. Above all, we must strive to protect and preserve the biosphere by living in a state of equilibrium rather than with the insatiable need for constant growth.

In bringing this book to a close, I cannot help but wonder where humanity is headed. The world has become a small place and our planet, floating in the immensity of space, is not unlike the Easter Island isolated in the vastness of the Pacific Ocean several centuries ago. And just as on Easter Island, the population is burgeoning, resources are diminishing, and the environment is deteriorating. I despair that we are about to repeat history. Nevertheless, I cling to the hope that we can reverse the situation before the last tree falls.

References

Information about historical figures is readily available from many sources so I do not always give references for them. Unless stated otherwise, I have used *Encyclopaedia Britannica, The Ultimate Reference Suite* (computer version), Chicago, 2003; the *Dictionary of Scientific Biography*, ed. Charles C. Gillispie (Scribner: New York, 1970–80); and occasionally Wikipedia.

acidrain.org, "Global Emission Trends." *Acid News* 2:2005. Online at www.acidrain.org/pages/publications/acidnews/2005/AN2-05.asp#global_emissions.

American Lung Association, *State of the Air: 2006.* Online at lungaction.org/reports/sota06exec_summ.html.

Air Quality Management District (AQMD), "The Southland's War on Smog: Fifty years of Progress toward Clean Air." 1997. Online at aqmd.gov/newsl/Archives/History/marchcov.html.

Audus, H., "Leading Options for the Capture of CO_2 at Power Stations." *Proceedings of the Fifth International Conference on Greenhouse Gas Control Technologies.* Cairns, Australia: CSIRO Publishing, 2000.

Blett, T., Geiser, L., and E. Porter, "Air Pollution-Related Lichen Monitoring in National Parks, Forests and Refuge." Denver: US Department of Interior and Department of Agriculture, 2003.

Bradsher, K. and D. Barbosa, "Pollution from Chinese Coal Casts a Global Shadow." *New York Times*, June 11, 2006.

British Petroleum, "Statistical Review of World Energy 2008." 2008. Online at www.bp.com/sectiongenericarticle.do?categoryId=9023784&contentId=7044480.

Brown, L.R., *Plan B 3.0: Mobilizing to Save Civilization.* New York: Norton, 2008.

Brown, M.H., *The Toxic Cloud.* New York: Harper & Row, 1987.

Brown, J. and M. Palacios, "The State of Urban Air in Canada." *Public Policy Sources* 85. Calgary: Fraser Institute, 2005.

Bryson, B., *A Short History of Nearly Everything.* Toronto: Anchor, 2003.

Cagin, S., and P. Dray, *Between Earth and Sky: How CFCs Changed our World and Endangered the Ozone Layer.* New York: Pantheon, 1993.

car-accidents.com, "Car Accident Statistics." Online at http://www.car-accidents.com/pages/stats.html.

Cahill, T., "Trans-Pacific Air Pollution is Worse than Was Suspected." 2000. Online at www.climateark.org/articles/2000/3rd/transpac.htm.

Canadian Encyclopedia, "Arctic Haze." 2008. Online at www.thecanadianencyclopedia.com.

Carley, L., and R. Freudenberger, *Understanding Automotive Emissions Control.* New York: HP Books, 1995.

Carson, R., *Silent Spring.* Cambridge: Riverside, 1962.

Canadian Council of the Ministers of Environment (CCME), *2003 Annual Progress Report on The Canada-Wide Acid Rain Strategy for Post-2000*. Ottawa: CCME, 2006.

Clean Air Initiative, *China Eyes Sulfur Dioxide Emissions Trading*. 2006. Online at www.cleanairnet.org/caiasia/1412/article-71096.html.

California Natural Gas Vehicle Partnership (CNGVP), "Reducing California's Petroleum Dependence." Online at www.cngvp.org/articles_wuebben.html.

Pearce, F., "Arctic to Lose All Summer Ice by 2100." *New Scientist*, December 4, 2002.

Commoner, B., P.W. Bartlett, H. Eisl, and K. Couchot, "Long-Range Air Transport of Dioxin from North American Sources to Ecologically Vulnerable Receptors in Nunavut, Arctic Canada." Montreal: Report for North American Commission for Environmental Cooperation, 2000.

Corcoran, E., "Trends in Energy: Cleaning Up Coal." *Scientific American* 264(5), May 1991.

Costanza, R. et al., "The Value of the World's Ecosystem Services and Natural Capital." *Nature* 387(15), 1997.

Canada Safety Council (CSC), "Traffic Accident Information." Online at www.safety-council.org.

Canadian Steel Producers Association (CPA), "Progress Report on the Environment for the Year 2000." 2001. Online at www.canadiansteel.ca.

de Villiers, M., *Windswept: The Story of Wind and Weather*. Toronto: McClelland & Stewart, 2006.

Diamond, J., *Collapse: How Societies Choose to Fail or Succeed*. New York: Penguin, 2005.

Dickson, D.R., and N. Quickert, "The Chemical Composition of Photochemical Air Pollution." *Photochemical Air Pollution: Formation, Transport and Effects*. Report 12. Ottawa: National Research Council of Canada (Associate Committee on Scientific Criteria for Environmental Quality), 1975.

US Department of Energy (DOE), "Environmental Benefits of Clean Coal Technologies." Topical Report 18 (April 2001). Online at www.netl.doe.gov/technologies/coalpower/cctc/topicalreports/pdfs/topical18.pdf.

US Department of Energy (DOE), "Alternative and Advanced Fuels website." 2008. Online at www.afdc.energy.gov/afdc/fuels/index.html.

Dotto, L., "Outbreak: The Climate Connection." *Globe and Mail*, 30 August, 2003.

Drake, F., *Global Warming: The Science of Climate Change*. London: Arnold, 2000.

Earthdive.com, "Doomsday Clock Moves Closer to Midnight." 2007. Online at www.earthdive.com/site/news/newsdetail.asp?changedate=true&changeyear=2007&id=1980.

European Academy of the Urban Environment (EAUE), "Malmo: Bo01, City of Tomorrow." 2001. Online at www.eaue.de/winuwd/187.html.

Economic Commission for Europe, "Strategies and Policies for Air Pollution Abatement: A Major Review Prepared Under the Convention on Long-Range Transboundary Air Pollution." Geneva/New York: United Nations, 1999. Online at www.htap.org.

Energy Information Administration (EIA), "Annual Energy Outlook 2003 with Projections to 2025." Darby, PA: Diane Publishing, 2003. Online at www.eia.doe.gov/oiaf/aeo/emission.html.

——, "World Carbon Dioxide Emissions from the Consumption and Flaring of Fossil Fuels: 1980–2005." 2005. Online at www.eia.doe.gov/iea/carbon.html.

——, "Germany: Environmental Issues." 2003. Online at www.eia.doe.gov/emeu/cabs/germe.html.

——, "Frequently Asked Questions: Renewable and Alternative Energy Sources." 2008. Online at tonto.eia.doe.gov/ask/renewables_faqs.asp.

——, "Coal Production and Number of Mines by State and Mine Type." 2008. Online at www.eia.doe.gov/cneaf/coal/page/acr/table1.html.

——, "US Coal Consumption by End Use Sector – 2006." 2006. Online at www.eia.doe.gov/cneaf/coal/page/acr/table26.html.

Elliott, S., "Mercury Showers in High Arctic under Scrutiny." *Environmental Science & Engineering* 16(4), 2003.

Emiliani, C., *Planet Earth: Cosmology, Geology, and the Evolution of Life and Environment.* New York: Cambridge University Press, 1991.

Environment Canada, *A State of the Environment Report: Canadian Perspectives on Air Pollution.* Ottawa: Environment Canada, 1990.

——, "Ambient Particulate Matter: An Overview." Hull, Quebec: Environment Canada, 1998a.

——, "The Canada-Wide Acid Rain Strategy for Post-2000." Ottawa: Environment Canada, 1998. Online at www.ec.gc.ca/acidrain/strat/strat_e.htm.

——, "The Stratospheric Ozone Primer." Ottawa: Environment Canada, 1999.

——, "Trucks and Air Emissions." Ottawa: Environment Canada, 2001.

——, "Summer Severe Weather." 2005. Online at www.mb.ec.gc.ca/air/summersevere/index.en.html.

——, "What's Being Done (about Acid Rain)?," 2003. Online at www.ec.gc.ca/acidrain/done-canada.html.

——, "Canadian Cities are Weather-Winners!" 2005. Online at www.on.ec.gc.ca/weather/winners/intro-e.html.

——, "Sulfur in Diesel Fuel Regulations." 2008. Online at www.ec.gc.ca/cleanair-airpur/Sulphur_in_Diesel_Fuel_Regulations-WS5B4D506F-1_En.htm.

——, "Acid Rain … and the Facts." 2005. Online at www.ec.gc.ca/acidrain/acidfact.html.

——, "Air Quality Health Index." 2008. Online at www.ec.gc.ca/cas-aqhi.

Environmental Protection Agency (EPA), "National Ambient Air Quality Standards." Online at www.epa.gov/air/criteria.html.

——, "EPA Fact Sheet: The National Ambient Air Monitoring Strategy." 2002. Online at http://www.epa.gov/ttnamti1/files/ambient/monitorstrat/fact.pdf.

——, "EPA Proposes Stronger Air Quality Standards for Lead." 2008a. Online at yosemite.epa.gov.

——, "Class I Ozone-Depleting Substances." 2008b. Online at www.epa.gov/ozone/science/ods/classone.html.

——, "Landfill Methane Outreach Program." 2008c. Online at www.epa.gov/landfill/overview.htm#converting.

Ertico, "Traffic Control Systems to Reduce Pollution (Athens, Greece)." Online at www.ertico.com/its_basi/succstor/trapocon.htm. — can't get this link to work.

European Environment Agency, 1999. *Air pollution in Athens: existing status and abatement practices.* Online at http://reports.eea.europa.eu/2599XXX/en/page018.html.

European Space Agency, "Record ozone loss during 2006 over South Pole." 2006. Online at: www.esa.int/esaCP/SEMQBOKKKSE_index_0.html.

Farman, J.C., B.G. Gardiner, and J.D. Shanklin, "Large Losses of Total Ozone in Antarctica Reveal Seasonal ClO_x/No_x Interaction." *Nature* 315(207), May 1985.

Farmers' Almanac: The 10 Worst Weather Cities. 2006a. Online at www.farmersalmanac. com/weather_chatter/2006/10/06/the-10-worst-weather-cities/.

Farmers' Almanac: The 10 Best Weather Cities, 2006b. Online at www.farmersalmanac. com/weather_chatter/2006/10/05/the-10-best-weather-cities/.

Farrand, Jr., J., *Weather*. New York: Stewart, Tabori & Chang, 1990.

Fisher, D.E., *Fire & Ice: The Greenhouse Effect, Ozone Depletion and Nuclear Winter*. New York: Harper & Row, 1990.

Flannery, T., *The Weather Makers: How We Are Changing the Climate and What It Means for Life on Earth*. Toronto: HarperCollins, 2005.

Francis, D., and H. Hengeveld, *Climate Change Digest: Extreme Weather and Climate Change*. Ottawa: Environment Canada, 1998.

Freeman, A. and J. Lewington, "All Clear in Central London." *Globe and Mail* March 1, 2003.

Freese, B., *Coal: A Human History*. New York: Penguin, 2003.

Gelbspan, R., *The Heat is On: The High Stakes Battle over Earth's Threatened Climate*. New York: Addison-Wesley, 1997.

Goodell, J., *Big Coal: The Dirty Secret behind America's Energy Future*. Boston: Houghton Mifflin, 2006.

Gore, A., *Earth in the Balance*. Boston: Houghton Mifflin, 1992.

Government Working Group (Canada), "Setting a Level for Sulphur in Gasoline and Diesel Fuel." Final report. Ottawa: Environment Canada, 1998.

Grady, W., *The Quiet Limit of the World: A Journey to the North Pole to Investigate Global Warming*. Toronto: Macfarlane Walter & Ross, 1997.

Gribbin, J., ed., *The Breathing Planet: A New Scientist Guide*. London: Basil Blackwell and New Scientist, 1986.

Hall, P. et al., *Effects of Acidic Deposition on Canada's Forests*. Ottawa: Natural Resources Canada, 1998.

Hansen, James, "Climate Tipping Points: The Threat to the Planet." Presentation at Illinois Wesleyan University, February 19, 2008. Online at http://www.columbia. edu/~jeh1/2008/illwesleyan_20080219.pdf.

Harte, J., C. Holdren, R. Schneider, and C. Shirley, *Toxics A to Z: A Guide to Everyday Pollution Hazards*. Berkeley: University of California Press, 1991.

Hengeveld, H.G., E. Bush, and P. Edwards, *Frequently Asked Questions about Climate Change Science*. Ottawa: Environment Canada, 2002.

Heverly, M., "Met One Instruments." Personal communication, August 4 and 20, 2003.

Hewings, J., "Air Quality Indices: A Review." Toronto: Pollution Probe, 2001.

Houghton, John, *Global Warming: The Complete Briefing*. Cambridge: Cambridge University Press, 1997.

Hotz, R.L., "Is Asia's Bad Air Stirring Storms in West?," *The Seattle Times*, March 6, 2007.

Harvard School of Public Health (HSPH), "Study Details Impact of Pollution on Public Health from Nine Older Fossil Fuel Power Plants in Illinois." Boston: HSPH, 2001. Online at www.hsph.harvard.edu/press/releases/press01032001.html.

International Joint Commission (IJC), "Air Quality Agreement: 2002 Progress Report." Washington: U.S. Environmental Protection Agency, 2002.

Interagency Monitoring of Protected Visual Environments (IMPROVE), "Deciview: A Standard Visibility Index." Newsletter 2(1). Fort Collins, CO: Air Resource Specialists, 1993.

Intergovernmental Panel on Climate Change (IPCC), *IPCC Second Assessment Report: A Report of the Intergovernmental Panel on Climate Change.* Geneva: IPCC, 1995.

——, *Climate Change 2001: The Scientific Basis: Contribution of Working Group I to the Third Assessment Report of the Intergovernmental Panel on Climate Change.* Cambridge: Cambridge University Press, 2001.

——, *Report of Working Group I, Summary for Policy Makers.* 2007. Online at www.ipcc.ch/pdf/assessment-report/ar4/wg1/ar4-wg1-spm.pdf

Iredale, W., "Polar Bears Drown as Ice Shelf Melts." *Times Online*, December 18, 2005. Online at www.timesonline.co.uk/tol/news/uk/article767459.ece.

Johnson, R.L., *Investigating the Ozone Hole.* Minneapolis: Lerner, 1993.

Kachan, D., *Oil Industry Subsidies for Dummies.* San Francisco: Cleantech, 2007. Online at www.cleantech.com/news/node/554.

Kahn, J. and J. Yardley, "As China Roars, Pollution Reaches Deadly Extremes." *New York Times*, August 26, 2007.

Keeling, C.D, and T.P. Whorf, "Atmospheric Carbon Dioxide Record from Mauna Loa: 1958–2004." Online at cdiac.ornl.gov/pub/ndp001/maunaloa.co2.

Keith, D.W., "Towards a Strategy for Implementing CO2 Capture and Storage in Canada." Ottawa: Environment Canada, 2002.

Kidd, J.S., and R.A. Kidd, *Into Thin Air: The Problem of Air Pollution.* New York: Facts on File, 1998.

Komp, R.J., *Practical Photovoltaics: Electricity from Solar Cells.* Ann Arbor: Aatec, 1995.

Lazaroff, C., "U.S. Legislature Votes to Withhold Information about Air Pollution." Environment News Service, June 22, 2000.

Leggett, J., *The Carbon War: Global Warming at the End of the Oil Era.* London: Penguin, 1999.

Liberal Party, *The Ontario Liberal Plan for Clean, Safe Communities that Work.* Pamphlet. Toronto, 2003.

Lomborg, B., *The Sceptical Environmentalist: Measuring the Real State of the World.* Cambridge: Cambridge University Press, 2001.

Lovelock, J., www.ecolo.org/lovelock.

Lynch, J., *The Weather.* Toronto: Firefly, 2002.

Marotte, B., "Exchange to Trade Emission Credits," *Globe and Mail*, July 13, 2006.

Max Planck Institut for Meteorology, "Investigation of the Global Distribution of Persistent Pollutants: The Grasshopper Effect." 2005. Online at http://www.mpimet.mpg.de/en/presse/pressemitteilungen/grashuepfer-effekt.html.

McIlroy, A., "Rural Trees Suffer more from Ozone, Study Finds." *Globe and Mail*, July 10, 2003.

McKibben, B., *The End of Nature*. New York: Random House, 1989.

Melillo et al., "Global Climate Change and Terrestrial Net Primary Production." *Nature* 363(234–240): 20 May 1993.

Meteorological Services Canada, "Ozone Balloon Research." 2004. Online at www.msc.ec.gc.ca/research/balloon.

Mihlar, Fazil, "Ethanol Anything but a Wonder Fuel." *National Post*, June 7, 2003.

Mintel International Group, "Automobile Purchase Process: US." North Adams, MA: Mindbrach, 2005. Online at www.mindbranch.com/listing/product/R560-1685.html.

Mitchell, G.J., *World on Fire: Saving an Endangered Earth*. New York: Charles Scribner's Sons, 1991.

Ministry of Environment (MOE), "Coal-Fired Electricity Generation in Ontario." PIBS 4016. Toronto: MOE, 2001.

——, "Ontario's Anti-Smog Action Plan: Progress through 2002." Toronto: Ontario Ministry of the Environment. Online at www.ene.gov.on.ca/envision/air/smog/asap2002.htm.

Monbiot, G., *Heat: How to Stop the Planet from Burning*. Toronto: Anchor, 2007.

Morgan, D., "Brazil's Biofuel Strategy Pays Off as Gas Prices Soar." *Washington Post*, June 18, 2005. Online at www.washingtonpost.com/wp-dyn/content/article/2005/06/17/AR2005061701440.html.

Nadis, S., and J.J. Mackenzie, *Car Trouble: How New Technology, Clean Fuels, and Creative Thinking Can Revive the Auto Industry and Save our Cities from Smog and Gridlock*. Boston: Beacon, 1993.

National Aerospace and Space Administration (NASA). "MODIS web." Online at modis.gsfc.nasa.gov/about.

National Highway Traffic Safety Administration, "Traffic Safety Facts 2006." Washington: NHTSA, 2006. Online at http://wwwnrd.nhtsa.dot.gov/Pubs/TSF2006FE.PDF.

National Mining Association. Personal communication with Leslie Coleman re. number of US coal-fired power plants, January 17, 2007.

National Oceanic and Atmospheric Administration (NOAA), "NOAA and Louisiana Scientists Say Gulf of Mexico 'Dead Zone' Could Be Largest Since Measurements Began in 1985." Washington: NOAA, 2007. Online at www.publicaffairs.noaa.gov/releases2007/jul07/noaa07-037.html.

——, "NOAA Celebrates 200 Years, Mauna Loa Monthly Mean Carbon Dioxide." Washington: NOAA, 2007. Online at www.celebrating200years.noaa.gov/datasets/mauna/image3b.html.

National Park Services (Air Resources Division), *Air Quality in the National Parks*, second edition. Washington: US Department of the Interior, 2002.

Natural History Museum, "Lichens: Silent Witnesses of Air Quality." London: Natural History Museum, 2004.

National Climatic Data Center (NCDC), *U.S. Climate Extremes Index*. Washington: NCDC, 2008. Online at www.ncdc.noaa.gov/oa/climate/research/cei/cei.html.

O'Connor, W.K., et al, "CO_2 Storage in Solid Form: A Study of the Direct Mineral Carbonation." *Proceedings of the Fifth International Conference on Greenhouse Gas Control Technologies*. Cairns, Australia: CSIRO, 2000.

Officer, C. and J. Page, *Tales of the Earth: Paroxysms and Perturbations of the Blue Planet*. New York: Oxford University Press, 1993.

Ontario College of Family Physicians, *The Health Impacts of Urban Sprawl: Air Pollution*. Toronto: OCFP, 2005. Online at www.ocfp.on.ca.

Pew Environment Group, *History of Fuel Economy: One Decade of Innovation, Two Decades of Inaction*. Online at www.pewfuelefficiency.org/docs/cafe_history.pdf, 2007.

Phillips, D., *The Climates of Canada*. Ottawa: Canadian Government Publishing Centre, 1990.

——, *The Day Niagara Falls Ran Dry! Canadian Weather Facts and Trivia*. Toronto: Key Porter, 1993.

Pollution Probe, *The Costs of the Car: A Preliminary Study of the Environmental and Social Costs Associated with Private Car Use in Ontario*. Toronto: Pollution Probe, 1991.

Pringle, L., *Rain of Troubles: The Science and Politics of Acid Rain*. New York: Macmillan, 1988.

Radio Free Europe, "Coal Mining Statistics." 2002. Online at wwwl.rfrerl.org/nca/features/2002/07/10072002152247.asp.

Ramanathan, V., et al., Warming trends in Asia amplified by brown cloud solar absorption Nature 448(575–578), 2007.

Recycling Council of Ontario (RCO), "Emission facts." Toronto: RCO, 2008.

Reid, K., "Advisories often Miss Bad Air Days." *St. Catharines Standard*, November 12, 2002.

Safe Kids USA, *Poisoning Fact Sheet*. Washington (DC): NSKC, 2004.

Sanford, J., "A Growing Concern: Planning to Cash in on the Ethanol Boom?" *Canadian Business* (90–98), 2006.

Sarma, K.M., "Protection of the Ozone Layer: A Success Story of UNEP." *Linkages Journal* 3(3), 28 July, 1998. Online at www.iisd.ca/journal/sarma.html.

Schaefer, V. J. and J.A. Day, *A Field Guide to the Atmosphere*. Boston: Houghton Mifflin, 1981.

Schlesinger, W.H., *Biogeochemistry: An Analysis of Global Change*. New York: Academic, 1997.

Seelye, K., "EPA To Issue Air Rules To Protect Park Vistas." *New York Times*, June 22, 2001.

Sherman, J., *Gasp: The Swift and Terrible Beauty of Air*. Emeryville, CA: Shoemaker & Hoard, 2004.

"Ocean 'Dead Zone' Alert," *Sidney Morning Herald*. March 29, 2004. Online at www.smh.com.au/articles/2004/03/29/1080544412449.html.

Simpson, S., "World on Brink of Severe Energy Shortage: Report." *Times Colonist* (Victoria), November 13, 2006.

Smith, D. M., "Recent Increase in the Length of the Melt Season of Perennial Arctic Sea Ice." *Geophysical Research Letters* 25(5) 1998.

Standing Senate Committee on Energy, the Environment and Natural Resources, *The Energy Emissions Crisis: A Viable Alternative*. Ottawa: Government of Canada, 1993.

Tammemagi, H., and D. Jackson, *Half Lives: An Introduction to Nuclear Technology in Canada*. Toronto: Oxford University Press, 2009.

Taylor, P.S., "Road to Nowhere: Canada's Traffic Snarl Costs our Economy Billions." *Canadian Business Online*, May–June 2006.

Tollefson, C., C. Rhone, and C. Rolfe, *Cleanair.ca: A Citizen's Action Guide*. Project of Environmental Law Centre, University of Victoria, British Columbia, Sierra Legal Defence Fund, 2000.

Turle, R. (Chief, Analysis and Air Quality Section, Environmental Technology Centre, Environment Canada). Personal communication, May 13, 2003.

Vital Climate Graphics, "Temperature and CO_2 over 400,000 years." Undated. Online at www.grida.no/climate/vital/02.htm.

United Nations Development Program, *Human Development Report*. New York: Oxford University Press, 1998.

United Nations Environment Program (UNEP), "GEO-2000: Global Environment Outlook." Online at www.unep.org/geo2000/english/0197.htm.

——, *North America's Environment: A 30-Year State of the Environment and Policy Retrospective*. Nairobi and Washington, DC: United Nations, 2002.

Von Dongen, M., "Clearing the Air on Dust and Pollen." *St. Catharines Standard*, August 22, 2002.

Walters, K., "Troubled Waters: Chapter 3." *St. Catharines Standard*, June 17, 2003.

Warrick, J., "Appalachia is Paying Price for White House Rule Change," *Washington Post*, August 17, 2004. Online at www.washingtonpost.com/ac2/wp-dyn/A6462-2004Aug16?language=printer.

White, E., "N.Y. Gives Ontario Pollution Black Eye," *St. Catharines Standard*, May 2, 2003.

Wolfe, C.W. et al, *Earth and Space Science*. Boston: Heath, 1966.

World Health Organization (WHO), "Air Pollution and Traffic Fatalities Info." Online at www.earth-policy.org/updates/update17.htm.

——, "One in Three Child Deaths in Europe Due to Environment." Online at www.euro.who.int/mediacentre/PR/2004/20040617_1.

Willis, S., *Vive le Kyoto: A Carbon Tax Primer*, HazMat Management, August/September 2006.

World Meteorological Organization and United Nations Environment Programme (WMO/UNEP), "Executive Summary—Scientific Assessment of Ozone Depletion: 2006." Online at ozone.unep.org/Assessment_Panels/SAP/Scientific_Assessment_2006_Exec_Summary.pdf.

Woods, A., "Harper Presses Alternative Accord to Kyoto." *Times Colonist* (Victoria), May 20, 2006.

World Health Organization and United Nations Environment Programme, *Urban Air Pollution in Megacities of the World*. Oxford: Blackwell, 1992.

Wright, A., "Coal Power Can Fuel This Century, Too." *Globe & Mail*, July 2, 2003.

Wright, R., *A Short History of Progress*, Toronto: Anansi, 2004.

Young, L.B., *Earth's Aura*. New York: Knopf, 1977.

——, *Sowing the Wind*. New York: Prentice Hall, 1990.

Ziock, H.-J., K.S. Lackner, and D.P. Harrison, "Zero Emission Coal Power: A New Concept." Paper Given at the First National Conference on Carbon Sequestration, Washington D.C., May 15–17, 2001.

Index